Alexander Black

Captain Kodak

A camera story

Alexander Black

Captain Kodak
A camera story

ISBN/EAN: 9783337082895

Printed in Europe, USA, Canada, Australia, Japan

Cover: Foto ©berggeist007 / pixelio.de

More available books at **www.hansebooks.com**

CAPTAIN KODAK

A CAMERA STORY

By

Alexander Black

Author of "Miss Jerry," "Miss America," "The Story of Ohio" etc.

With Photographic Illustrations by the Author

BOSTON

LOTHROP PUBLISHING COMPANY

Plimpton Press
H. M. PLIMPTON & CO., PRINTERS & BINDERS,
NORWOOD, MASS., U.S.A.

TYPOGRAPHY BY J. S. CUSHING & CO., NORWOOD, MASS., U.S.A.

To My Boys

CARL AND MALCOLM

A WORD AT THE BEGINNING

IN the olden days people peered about in the world for the fountain of perpetual youth. Nowadays they are wiser. They find a hobby — an enthusiasm, by which the old are made young and the young are made wise and happy.

This is the story of the camera hobby; of an amateur photographer and his chums; of a boy's adventures in the company of his camera; of a camera club and the old and young brought together by the influence of a common interest; of journeys in search of the picturesque; of problems, struggles, and surprises.

The pictures are not by any means always intended to show my readers how photographs should be made, but rather to suggest the interest of familiar and accessible things, and that the best thing about a photograph is not always the thing we wished or expected to put in it.

For assistance in the form of pictures at the Montauk camp I am indebted to my friend, Mr. Walter Hammitt.

A. B.

CONTENTS

CAPTAIN KODAK.

I.

THE COMING OF THE CAMERA.

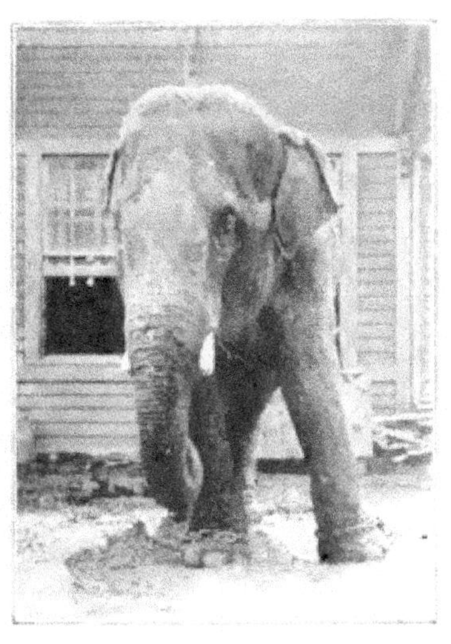

ON the day when the circus came to Hazenfield, one of the elephants broke loose and strolled up Main Street; and when they chased him he knocked down three lamp-posts, the stone boy on the drinking-fountain, upset a trolley car, broke the insurance company's sign, smashed the helmet of Policeman Ryan, and fell into a hole in front of the barber's.

There never had been so much excitement in Hazenfield, and the motorman, Policeman Ryan, and the barber hope there never will be again.

When it was all over, that is to say, when they got the elephant out of the hole, which you must know was no easy matter, and Hazenfield had quieted down again, there were many comments on the incident.

"I never expected an elephant," said the motorman.

"I'm glad it wasn't your head," said Policeman Ryan's wife, when she saw the helmet.

"I thought he was coming in to get shaved!" said the barber.

Allan Hartel, the Doctor's son, said, "If I'd only had a camera!"

Allan recalled how Main Street cleared, or tried to clear, when the elephant was first discovered; and the way the elephant swung his trunk, and dropped into a hobbling trot before he struck the trolley car. He recalled the frantic movement of the motorman as he caught sight of the big, lumbering beast at the corner.

"If I'd only had a camera!"

He recalled the brave way that Policeman Ryan stepped out into the street, waving his club, and the way he dodged when the elephant swung at him with his trunk.

"If I'd only had a camera!"

He recalled the way the elephant dropped on his knees in the hole. He recalled the funny wrinkling of the elephant's hind legs as if he had on a pair of trousers that were too large for him.

"If I'd only had a camera!"

I suppose that the way he felt about this elephant affair had a good deal to do with the fact that after that Allan always liked so much to photograph elephants. But I must not get ahead of my story.

To properly go on with the story I must tell you that about six weeks after the elephant got himself in a hole, and the circus people, with derrick and tackle, got him out again (you never saw an elephant more truly ashamed of himself than that elephant), Little

McConnell saw Allan Hartel come out of the express office with a package.

Now you, reader, will guess at once that this was a camera, but McConnell had no suspicion of this fact.

"Hello!" called McConnell, "what have you got there?"

McConnell was thirteen, two years younger than Allan. He was called Little McConnell to distinguish him from his brother, who was called Big McConnell. It would be hard to say why no one ever called him Percy — his first name. Even Allan always called him simply McConnell. He was the kind of boy, somehow, that you always call by his last name and never know why.

McConnell and Allan had been chums for a long time, and McConnell certainly should have known what was in the bundle had he not been up to Greenby visiting his aunt for two weeks, and had not Allan kept a certain little enterprise a secret from everybody before that. But when Allan said, "Guess," he was much puzzled for a moment. Then he made the most successful guess he ever had made in his life.

"Not a camera?" he exclaimed.

"Yes," admitted Allan.

"When did you buy it?" McConnell felt as if he must have been left out of Allan's confidence somehow.

"I didn't buy it," Allan replied.

"Then who gave it to you — your father?"

"It wasn't given to me," returned Allan.

"Well," said McConnell, a little annoyed, "that is just a trick. You'd have to buy it or have it given to you — wouldn't you?"

"No," said Allan; "there's another way."

"Oh, yes — you could find it."

"Then there is still another way," Allan insisted.

"You don't mean to steal it, do you?"

"No," said Allan; "there is one other way."

"I give it up," said McConnell. "That conundrum beats me," and he went over the thing on his fingers: "buy it, have it given to you, find it, steal it, — what else is there?"

"Win it," said Allan.

McConnell laughed. "Cheney says 'win' when he means steal."

"I can't help that," insisted Allan; "I did win it."

"How? What was the game?"

"It wasn't a game. I wrote a composition. There were a lot of prizes. One of them was a camera."

"You always were lucky," said McConnell. Then to show that he wasn't envious, he added: "I'm glad you did win it. I was thinking the other day that everybody seemed to have a camera except us. Is it a 'press the button'?"

"It's both. You can press the button or stand it on legs, either one. It hasn't any legs, now. They come separately. I don't believe I'll care much for them. I can rest it on something."

"Yes," McConnell assented; "when they're on legs they sometimes get broken when some one kicks against one of the legs. Let's see, what is it they call the legs?"

"Do you mean tripod?"

"Yes, that's it, tripod. I wonder why it isn't tri-*ped*," mused McConnell, as they continued their walk toward Allan's house. "We say biped and quadruped for two legs and four legs."

"McConnell and Allan had been chums for a long time."

Allan could not explain; and he was thinking about the camera. "Don't you want to help me fix up a dark room out in the stable?"

"That's just what I do want," exclaimed McConnell. "I want to learn the ropes. You see, I think that when Bill hears about your having a camera he'll help me to get one somehow. It seems to me," McConnell continued enthusiastically, "I'd almost swap my wheel for one!"

Allan was thinking about the dark room. "Jo Bassett has his in the kitchen. I mean he develops there at night, and Owen has his in the attic. I wanted father to let me have the little place by his office, you know, where all the bottles are, but he said, No, sir! I'd have to doctor my plates where he wasn't doctoring his patients, for he didn't want either the plates or the patients to get the wrong doses."

The boys laughed.

"Is the stuff dangerous that they put on the plates?" asked McConnell.

"I guess not," answered Allan, "unless you drink it. Father says there are two sides to a person, the inside and the outside, and he says we mustn't use things on the wrong side. He's going to help me about the bottles."

"But you must take the pictures first," said McConnell. He was impatient to see the camera, and to have it aimed at something. "Couldn't we—couldn't you take something to-day?"

"It's too late now," said Allan, regretfully. "We need a lot of light, and there's scarcely any left. But we'll get everything ready, so far as we can, for to-morrow."

When they reached Allan's house the Doctor was

just getting into his carriage at the door. "Hello!" he called; "so it has come, Allan?"

"Yes, sir," and Allan swung his package in the air.

"Good!" exclaimed the Doctor. "I shall want to see it when I get back."

The boys made short work of the bundle when they reached indoors. Wrapped in strong paper and nestling in "excelsior" was the shiny, leather-covered box, with holes, and buttons, and levers, and gauges, — a mysterious box, which the boys proceeded to examine from its six sides with great reverence.

With the aid of the printed instructions, and what knowledge the boys had acquired f[illegible] seeing other Hazenfield cameras (especially Owen [illegible]nt's), the mysteries began one by one to seem less mysterious. It was great fun to watch the images of the room, of the window, of the street, in the little "fi[illegible]r."

"Isn't the picture going to be any bigger than that?" asked McConnell, in a disappointed [illegible].

"Oh, yes," said Allan; "that is only [illegible] show where the picture will come on the plate back here. It's only a miniature of the real picture."

"And it isn't upside down, either," remarked McConnell, peering into the little opening at the top of the box.

"Somebody told me," said Allan, "that was because there was a little piece of looking-glass on the inside that twisted the thing around."

Presently they found that by opening a lid and looking through the box from the back the real image from the lens fell on the "focussing glass," this time upside down.

McConnell laughed. "That always seems so funny." He twisted his head in an effort to get a natural view of the room on the glass. Then he ran across the room and stood on his head against the wall.

"Do I look right side up now?" he demanded of Allan.

"Yes," laughed Allan, peering into the box. "You look right side up, but you don't look very natural."

"Suppose you turned the camera upside down," suggested McConnell, coming back.

Allan laughed again. "I'm afraid that wouldn't do any good," and he turned the camera to show McConnell that the picture was still hopelessly inverted.

McConnell thought that he liked the "finder" picture better. "It's too bad," he said, "that it isn't bigger."

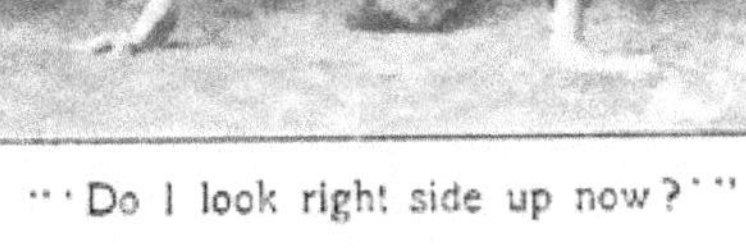

"'Do I look right side up now?'"

Allan had been reading about cameras. "There are special cameras," he said, "that have finders on top as large as the focussing glasses at the back."

McConnell thought that he would like one of that sort.

"What's the use?" asked Allan. "The little

finder tells you just what you are going to get. It's the picture boiled down — well!" Allan shook the box. "I hope something hasn't broken already." A rattling sound came from the inside of the box.

"Maybe it was broken in the express," ventured McConnell. But investigation proved that the rattling sound was produced by a loose screw under the front cover of the box, which the directions showed was to be used when the camera was placed on legs.

Having opened the front of the box to make this investigation, Allan was now able to closely examine the lens.

"That's the diaphragm," said Allan, pointing to the disk of metal protruding from the barrel of the lens.

"The what?"

"Diaphragm," repeated Allan.

"How do you spell it?"

"I don't think I can spell it. What do you always want to spell things for? It begins with a d-i-a and then gets all mixed up — ho, here it is in the directions, if you must spell it — 'd-i-a-p-h-r-a-g-m.'"

"What does it do?"

Allan was turning the disk. "Look," he said; and they saw that the diaphragm had thre[illegible]s in it, and that any one of these holes [illegible] rought opposite the centre of the lens.

"I don't see how anything could possibly get through that!" exclaimed McConnell, staring at the smallest opening.

"Why," said Allan, "Owen says you can photograph through a pinhole — *with* a pinhole, I think he said."

"He didn't mean without a lens, did he?" demanded McConnell, incredulously.

"That's an old trick, McConnell," said Dr. Hartel in the doorway. "I photographed with a pinhole when I was a lad."

"You did!" cried Allan. "You never told me about it."

"I don't see how the picture ever squeezes through," said McConnell.

"Light is wonderful," mused Allan, prying further into the box.

"Everything in nature is wonderful," said the Doctor "when you come to know about it. Your lens is wonderful, but not more wonderful than the hole among the leaves of a tree that photographs the sun on the ground underneath. It isn't any more wonderful than the way the plate catches and keeps the image."

"The plate!" repeated Allan. "I had forgotten about that! We can't make pictures unless we have something to make them on."

"I suppose you can get them at the photographer's, can't you?" asked the Doctor, examining the camera.

"Wincher's stationery store sells cameras," said [illegible] "and I guess they sell plates too."

[illegible] matter associated with the camera had an [illegible] erest for the boys that day. McConnell came around in the evening after Allan had run down to the stationer's to get a package of plates.

"Open by ruby light only," read Allan on the box.

"Yes," said the Doctor, "you'll have to think about your dark room."

"The dark room!" This seemed like one of the most interesting things about the whole affair.

"Though the box might have said, 'by ruby light or no light,'" replied the Doctor. "There is no objection to your opening it by no light if you want to."

"But we couldn't see," protested McConnell.

"You could feel, though," the Doctor explained. "An old photographer told me that he always preferred to load his plate-holders in the dark. He trusted his touch with no light more than he did with a weak red light with which he sometimes let his eyes deceive him."

"Deceive him how?" asked Allan.

"By letting him get a plate wrong side up."

"Oh!" said Allan. He hadn't thought to consider that the plates had a right and a wrong side.

"When you come to open your box," — then the Doctor paused a moment. "Suppose, boys, that we go and load the plate-holders. We'll go up to your room, Allan."

"But how about the ruby light?"

"Oh, we shall soon fix that. Where is your bicycle lamp?"

Allan fetched the well-worn headlight of his wheel, and when it was lighted, the boys remarked that the side glasses were a rich red.

"But what shall we do with the front glass?" and Allan struggled to think of some way to color the front glass.

"Wait a moment," said the Doctor. "You will get a regular ruby lamp if you need one, but I think I can show you an emergency method of using any lamp of this sort." He found a piece of reddish powder-wrapper in his chemical closet, and this he fastened over the front of the lantern; then taking a larger sheet of manila paper he made a cylinder of this about the size of an ordinary Chinese lantern.

"That," he said, "is a safety shield to keep out any rays of white light that may escape from any of these smaller ventilating holes of the lamp." The Doctor placed the lamp inside the shield. "Yellow paper is the next best shield to red. They got along with yellow light when photographic plates were less sensitive. Now they often use both yellow and red glass in combination. Well, I guess we are ready to load up."

Allan led the procession up to his room, carrying the plate-holders — there were four of these — and the lamp. McConnell came next with the manila paper shield, and the Doctor followed in the rear with the box of plates. On the way the procession met Mrs. Hartel who had been putting little Ellen to bed.

"What is this strange procession?" she cried.

"This is the kodak contingent," laughed the Doctor; "a company of kodakers just going into camp."

"And you, Allan, are you the captain?" asked Mrs. Hartel.

"Yes," the Doctor replied for him, "he is the captain — Captain Kodak; that is quite a good name for him now."

"Well," Mrs. Hartel called after them, "I hope you will always preserve good order in your camps — and especially great cleanliness. You know what I mean by that, Harry," Mrs. Hartel said to the Doctor. "I don't want any chemicals on the bed-spread."

"Oh, we're going out to the stable to do that," Allan called back.

"To do what? — put chemicals on the bed-spread?"

"No, no!" expostulated Allan, at the door of his room, — "I mean to use the chemicals."

They cleared a little table in Allan's room and placed the lamp in the centre of it, with the yellow paper shield in position. A soft, yellowish light filled the room and made the three faces look strangely unusual.

"This makes me think of a conspiracy," said Allan.

"Or three robbers in a cave," said McConnell.

"Now, you understand, boys, that I don't really know very much about photography," said the Doctor. "When I was studying medicine I had a room-mate who was a photographic crank, and I once saw him do something of this sort, though he used a small stable lantern with a red bandanna handkerchief tied about it. This ought to be much safer, and it needs to be, for plates are much more sensitive, even to red and yellow light, than they used to be. I suppose that some day they will make photographic plates so sensitive that we shall have to develop them absolutely in the dark."

"That would be harder than loading them in the dark, wouldn't it?"

"Decidedly harder. Now, boys, let us get out the plates. Probably I shall do something that I shouldn't do, and you will learn afterward not to do it. But I am better than no help at all, am I not?" the Doctor added laughingly.

"Yes, indeed!" Allan admitted.

The Doctor had used the point of his knife in cutting through the paper in the bottom of the box. Then they found that the plates were hidden in three boxes, one within the other.

"We must not expose the plates too long even to this faint light," the Doctor remarked, as he opened one of the plate-holders. Then he took out one of the plates and showed the boys that the plate was coated on one side with a yellowish substance; then, still keeping it in shadow, he let each of the boys feel both surfaces. The coated surface had a smooth feeling.

"The plain glass side feels sticky, doesn't it?" said McConnell.

"In the dark," said the Doctor, "you can easily tell the difference, though you should always feel the plates near the margin, because the moisture of the fingers may leave a stain that will afterward appear in the developed plate. My chum once photographed me sitting by the window, and a finger mark which fell on my face — that is to say, on which my face fell, for I think the carelessness was before the picture was taken — made me look like a very disreputable citizen indeed. My chum said I looked like a surprised pirate. Now, if you know how a surprised pirate looks, you can fancy my appearance. But usually you won't need to feel the plates to place them properly, for each maker packs his plates in a certain way. This maker packs them face to face. That, I should judge, is the usual way, now. And here goes for the plate-holders."

When the plate-holders each had their two plates in position, back to back, the Doctor said it would be well to remember that the four remaining plates in the package must not be confused with the others when the time came for developing or changing plates. But, he added, "I know well enough that you will have to make all these mistakes to know how to avoid them."

"Did you do that with your prescriptions, father?" asked Allan, with a grin that was not concealed even by the half darkness.

What the Doctor might have answered it is hard to say, for just then Mrs. Hartel knocked at the door to say that Owen had come. In fact, Owen was then at the door.

"Come in," called the Doctor, the plate box being safely closed.

"Hello, Owen," shouted Allan from behind the lantern. "I didn't go after you because I thought it was your night at the Choral."

"There wasn't any meeting to-night," Owen said, "and I just happened to hear from Cheney that you had a camera. What is it?"

"A Wizard," said Allan, "and a little beauty. I wish it was daylight. I hate to wait until to-morrow."

"What kind of a lamp is that?" asked Owen, puzzled by the object on the table.

"That," replied the Doctor, smiling at the group of boys, "is the famous Hartel Adjustable Lamp."

Owen saw the joke.

"I suppose we'll fix up something better in the stable," said Allan.

"In the stable?" Owen looked interested. "That's a good idea. Won't you let me see your Wizard?"

They all trooped downstairs again. "Here come the kodakers!" cried McConnell. There they found Mrs. Hartel and Edith Coles, Allan's cousin, studying the camera by the sitting-room lamp. Edith was an orphan niece of Mrs. Hartel who had been a member of the Hartel household for six years. She was now of about Allan's age, and always was much interested in everything that Allan did. Returning from the home of a girl friend where she had been spending the afternoon and early evening, she was as much delighted over the camera as if it were some good fortune of her own.

"'And is it all loaded and ready?' asked Edith"

"I want to be a kodaker, too!" she exclaimed laughingly in response to McConnell's jubilant announcement.

"I guess Allan will let you join his company," the Doctor said.

Owen was called upon as the most experienced in new cameras to tell Edith and the rest all about the Wizard; to explain the focussing scale, which Dr. Hartel said Captain Kodak really should call a "range-finder"; to point out features of the shutter, through which the picture could jump in the fraction of a second, or which could be set so that a long exposure might be given when there was not sufficient ligint for a "snap shot"; to show the action of the slides in the plate-holders, the use of the diaphragm, and more other things about the camera than you would have supposed could be said about a box so small and innocent looking.

"And is it all loaded and ready?" asked Edith, looking down at Allan and McConnell, who were bending over the camera in some new investigation.

"Yes," said Allan.

"It is a pity not to be able to try it now in some way."

"Edith," remonstrated Mrs. Hartel, "you are always impatient."

"Well," said the Doctor, "I guess they all are — I think I am myself. The only difference is that Edith speaks out."

"You could make a flash light," Owen suggested.

At this moment the clatter of a bell could be heard in the adjacent street and some one ran rapidly past the house.

"A fire!" shouted McConnell

There was a pause during which every one listened breathlessly. Allan and McConnell were already at the gate. "It *is* a fire!" Allan reported in a moment, "over by the East Church."

"There is something to photograph!" exclaimed the Doctor.

"Could I?" cried Allan, with an appealing look to Owen, — "at night?"

"Why, I should think night was a good time to photograph fire," Edith declared.

"It has been done," Owen admitted.

"I'll try it!" Allan caught up the camera. "Won't you come, Owen, and help?"

They all were at the door in a moment.

"Allan!" called the Doctor. "You've forgotten your hat!"

"Be careful where you go," warned Mrs. Hartel, as she pressed the hat on the boy's head with a motherly firmness.

II.

THE FIRE.

EDITH at the gate could see the three boys running in the direction of the red light in the sky. Allan, in the lead, was hugging the camera under his arm. There was a sound of many feet, a murmur of excitement in the air, and distant hoarse shouts. A huge roll of black smoke drifted off to the north.

"I believe it's the factory," said the Doctor, at Edith's shoulder. "Let us go too, Edith."

Edith did not wait for a second invitation. She had been longing to follow the boys, and had hurried upstairs for her hat and was again in the hall before the Doctor had reached for his cane.

"Margaret," called the Doctor to Mrs. Hartel, "perhaps you wish me to take you."

"No, indeed," said Mrs. Hartel. "I'm afraid I

don't appreciate the fun of fires. I had rather have all of you tell me about it."

"We shall!" cried Edith from the walk.

It was as the Doctor expected. Flames had broken out in the southern wing of the factory. The eastern windows on the first floor of the wing showed an orange-red glare that made Edith think of the reflected light on window-panes when the sun is going down. The flames evidently had passed through to the second floor and were creeping eastward, though the dense masses of smoke made it difficult at times to tell precisely what parts of the building were actually burning.

The Hazenfield engines were hard at work. The ground trembled with the thump, thump, thump of the steam pumps, the black figures of the firemen scurried this way and that with many confused shouts, while the inky line of the hose twisted its way to the wing door of the factory.

It was at the wing door that some men were carrying out certain heavy cans which they placed at some distance from the burning building. These men were dripping with the hose water, the light of the flames glittering on their clothes and faces.

"Why don't they pour the water through the windows?" cried Edith.

"Because they know they can't save the wing," replied the Doctor; "they don't want to increase the draught by breaking the windows before the heat destroys the glass, and they are fighting indoors to keep the flames from spreading to the main building."

Almost as the Doctor spoke four of the upper story windows blew out, a rush of flame following and mounting high over the roof.

"There!" exclaimed Edith, "they must pour in the water now!"

"Why, you almost seem glad, Edith," said the Doctor.

"Well," pleaded Edith, "it seems so foolish not to pour the water where most of the fire is."

Two streams of water now leaped up to the open windows and sizzled and snorted under the blazing eaves. The flames greeted the serpents of water with a howl of rage and defiance, and fresh clouds of smoke arose at the places where they fought together.

"I wonder where the boys are?" queried Edith.

The Doctor had been wondering the same thing.

In a great circle about the burning factory were the faces of the spectators gleaming in the firelight. The stillness of the crowd was astonishing. The crackle of the flames could be heard with a strange distinctness, and the hoarse voice of the engine foreman sounded clear above all other voices. Only when the window-glass fell out or some other fresh event of this kind happened, did the crowd make a noticeable sound. Then there would be a general murmur running completely around the circle.

If the Doctor and Edith could have seen more distinctly in the uncertain light the embankment to the south of the burning wing, where the crowd was thinnest, they would have discovered Allan and his two companions grouped closely in earnest consultation.

As he had been running to the fire it had seemed very odd and foolish to Allan to be carrying the camera, and when they actually reached the scene of the fire the carrying of the camera seemed even more foolish than before. Yet it certainly made the whole affair seem more like an adventure to have the camera along. It seemed like going armed.

"Phew!" ejaculated McConnell, "it's going to be a whopper! It's all blazing inside! Come over here!"

"Let's go around to where the engines are," suggested Allan.

"Here's a good place!" Owen called, pointing to the embankment.

The three boys clambered up the embankment in an excitement which only a fire can call out.

Allan's first thought was of the camera. "Do you think I might try a snap?" he asked Owen.

"I should try one snap," Owen suggested, "just as an experiment, and then try one or two seconds' time."

Allan fixed his "range-finder," as Doctor Hartel had called it, for the full distance, which made the focus accurate at fifty feet and beyond.

"Wait a moment!" cried McConnell, "it isn't blazing high now."

Allan was locating the factory in his "finder." Nothing but spots of fire were visible there. McConnell was eager for a glimpse of the little picture.

Presently Allan said, "I guess I'll snap it now, Owen!" and he pressed the trigger.

"That'll be great!" McConnell exclaimed.

"Did you draw the slide?" Owen asked.

Allan looked stupid. "No," he confessed, "I didn't." Then he opened the box, drew the slide that hid a plate, set his shutter again, Owen stooped forward to see that the front opening was uncovered, and Allan pressed the trigger once more.

"This time I guess we caught it!" Allan said.

"'Wait a moment!' cried McConnell."

Owen now advised that they rest the camera on a large stone for making the "time exposure," and he assisted Allan in setting the shutter so that the exposure could be made by opening and closing the sliding front of the box. Each moment the building became more brilliantly lighted. The flames had stolen across the end of the wing from east to west, and when Allan opened the little door for two seconds — McConnell quickly counted five in the same space of time — the main floor was more than half swept by the fire.

The efforts of the firemen to keep the fire in the wing seemed likely to succeed, though they could not have succeeded had there been any wind, particularly had there been a wind from the south. It was exciting to watch the battle between fire and water — the fire leaping blindly hither and thither like a wild beast; the water guided by skilful men who stood at their posts with hats pulled low to keep the blistering heat out of their faces.

While the boys were preparing for a third shot, the flames came streaming through a hole in the roof near the cornice, and fluttered like a great yellow banner thirty or forty feet long.

"Now!" screamed McConnell in great excitement, "there's a picture for you!"

Allan pressed the trigger, and not a moment too soon, for a stream of water struck the opening in the roof, and the great golden banner of fire shrank rapidly and finally disappeared in clouds of steam and smoke. The boys found it hard to watch the fire and not wish that the firemen would place the streams in some spot they seemed to have overlooked. It was like watching a man paint a fence or hoe a garden.

There were places which up to the last moment seemed likely to be forgotten altogether.

This was particularly true of a little river of fire in the cornice which slowly crept along until it seemed on the point of reaching the main building.

"I wonder why they don't put out that place in the cornice," Allan said impatiently; "I'm sure they don't see that."

Owen and McConnell had both noticed this stealthy movement of the fire.

"I almost feel like going over and telling them," said McConnell. "If they don't hurry it'll surely catch the main roof."

Then swish! came No. 2 engine's hose, and the little stream of fire instantly disappeared.

"Probably they know what they are about after all," admitted Allan.

"The spots of fire."

"The flames came streaming through a hole in the roof."

Owen laughed. "But I think, McConnell, you ought to go over and give them the advantage of your advice."

All three were sitting on the edge of the embankment watching the waning fire, when a voice in the darkness called, "Hello, boys!"

It was Dr. Hartel and Edith. "We have been looking for you everywhere," said the Doctor. "We watched for three boys in a bunch."

"Yes," laughed Edith, "there were different size bunches, and when we saw a bunch of three — "

"With a black box," put in the Doctor.

"— with a black box, we knew it was you."

"We have made three pictures of the fire," said Allan.

"You mean three exposures, don't you, Allan?" the Doctor asked, with his teasing smile. "Better wait until after the developing before talking about pictures."

"Anyway," persisted McConnell, "it was aimed right, and I heard it click."

"And I saw that the front door was open," laughed Owen.

"And the slide out," added Allan.

"Of course," said Owen, "I don't suppose there will be much on the plates but the flames."

"It is getting chilly, Edith," said the Doctor, after a time. "I guess that you and I had better stroll home. They have the thing under control now. Don't stay too long, boys."

The Doctor had not gone far when Allan came running after them to say that Owen had suggested using some of his developer for the plates, and that he was to bring his dark-room lamp with him. "We are going to begin up to the coach-house to-night."

"To-night?" repeated the Doctor.

"Yes," returned Allan, "it's only half-past nine, and Owen says we can do it all in an hour."

And so, after waiting about fifteen minutes longer, until the fire had dwindled to a point at which the fire chief saw only an hour's work ahead of him before leaving the blackened wing of the factory to the care of the watchman, the boys started for Owen's and were not long in starting back eagerly for the coach-house.

III.

UNDER THE RED LAMP.

ABOVE the stalls in Dr. Hartel's stable were three rooms, in one of which a coachman used to sleep in the days when the place had been used by Judge Solling. The two other rooms were only partly finished. In one of these was a sink with running water, which had long been marked in Allan's fancy as the focal point of the dark room.

"We can't fix anything here to-night," said Owen.

"Of course not," admitted Allan.

Owen had carried over two trays, "one for developing and the other for fixing," and at his suggestion Allan procured an "agate iron" tray from the kitchen to wash the plates in. "Mind you fetch it back!" said Nora.

"Is it dark enough here?" asked Allan, turning to the back windows.

"Yes," answered Owen; "but in the daytime you would have to cover up the windows in some way, and keep the daylight in the front rooms from getting in around these doors."

Meanwhile Owen, in the red light of his lamp, was fussing with two bottles, a proceeding which excited the greatest interest on the part of the two other boys. Allan often had seen his father make chemical experiments, and he had seen Owen develop once before; but he was not a photographer then, and had not watched each motion with the same feeling of concern and anticipation.

"I forgot my graduate," Owen complained.

"Shall I get one of father's?" asked Allan.

"No, I can guess the amount pretty well in this old glass."

Owen poured from each of his bottles, and then added water from the tap, inspected the trays critically, and turned the flame of the lamp a little lower. "Now," he said, "we are all ready."

"Ship ahoy!" came a voice from the stable stairs. It was the Doctor.

"Can't we come?" That was Edith.

"Yes, yes! Come right up!" shouted Allan, running to the head of the stairs to pilot the newcomers, "though I don't know where you are going to sit — we haven't any chairs."

"Oh, we shan't mind that!" said the Doctor.

"We are just ready," said Owen.

Allan thought it was good of Owen to say "we," for he himself had taken but small part in the important preparations.

"I hope this won't make you nervous, Owen," the Doctor said. "I don't know that I should want to perform an operation with so many onlookers."

"I may not do the right thing," Owen confessed; "but I only know how to do the one thing, anyway, and that is to pour on the developer and let the thing go."

The Doctor laughed quietly. "I see," he said, "you administer the medicine and let nature do the rest. After all, that is the about the most any of us can do."

"Now," asked Allan, "do you want one of the plates?" He had been standing with the plate-holders in his hand.

"Yes," Owen answered, "we're all ready."

They opened a holder and took out one of the plates. Owen placed the plate in one of the trays, poured his developing mixture over it, and began gently to rock the tray, the spectators crowding about him in a semicircle.

"Of course," said Owen, "it may be a long time coming up." Presently he added, "It may be very much under-exposed, you know."

At the end of five minutes the plate remained obstinately free from any sign of an image.

"I don't see a thing," said Edith.

"But, Edith!" expostulated Allan, "it sometimes takes a long while."

"I think I have done the thing right," murmured Owen, in perplexity. Then he suddenly turned to Allan. "Say, which one of the plates is this?"

Allan's face took on a queer look in the red light. "I don't know," he answered blankly. It had not occurred to him before. "I know we didn't double any."

"I hope not," Owen interposed anxiously.

"But I forgot to turn the holder-slides over to show which had been exposed."

"Well," said Owen, "we'll have to try them until we find the three exposed ones. You had the four holders in the box?"

"Yes," asserted Allan, "eight plates."

At the end of another five minutes, Owen said, "I'm sure this can't be one of the exposed plates," and, with a last flip of the developer, he took the plate out and placed it on the shelf over the sink.

"And can't you use it again?" asked Edith, sympathetically.

Owen shook his head. "No. *That* one's done for. Now," he added, "let us try one of the others."

"I'm sure this is one," said Allan, repentantly. He felt as if he had made a bad blunder on his first photographic expedition. "I remember now which way I carried the holders," and he handed Owen another plate.

The semicircle of faces drew about Owen's shoulders again. It seemed a long while to wait, while Owen rocked the tray affectionately, as if it were a cradle, and Allan's spirits had begun to fall, when McConnell cried, "There's something!"

"It's just a spot on the plate," said Edith.

Owen shook the tray as if to dislodge something from the plate. "Why, there are several of them!" he cried. "They must be the windows of the factory!"

"Hooray!" shouted McConnell.

"Black windows?" asked Edith, perplexed.

"It is the fire," said the doctor. "Light objects make a black image on the plate. That is why they can print from the negative. Daguerre made a positive — a natural image — on a metal plate which could not be duplicated. That was a daguerreotype. The English inventor of photography made negatives first on paper and then on glass. These could be used for making any number of positives or prints."

"I see," said Edith, her eyes on the plate. "Is anything more coming, Owen?"

"Yes," said McConnell, "more little spots, and they are getting blacker. I think I see some flames shooting up."

But Owen did not seem very sanguine. "It doesn't seem to come out very well," he mused. "I guess this one is the snap shot."

"Then there will be more on the two others!" cried Allan, hopefully.

They all were greatly interested to think that the other plates might have more on them. Owen's guess proved quite correct. The other plates from the same holder came up in a much shorter time.

"Why, yes!" Edith exclaimed. "You can see the window-sash plainly, and the fire is spreading in this one."

"You are right, Edith," said the Doctor. "There are flames in several more of the windows here than in the other plates. And I can see faint outlines of the building here and there — and what looks like a stream of water, lighted up by the fire, in another place."

"And so we have something after all!" Allan said, jubilantly.

The third plate displayed the fire at its worst, when the flames broke through the roof, and they were all watching the growth of the image under the soft swish of the developer when a sharp rap sounded on the stable door at the foot of the stairs.

"A call, uncle," said Edith, resentfully. "They always want you when we want you, don't they?"

The Doctor went to the door, and he could be heard talking in a low tone for some moments. Then he said, "Good night!" and came up again quickly.

"Didn't you have to go?" asked Edith.

"Here's something extraordinary, Allan!" the Doctor exclaimed; "the factory company wants your negatives!"

"My negatives!" Allan looked amazed.

"The fire pictures?" asked McConnell, staring at the Doctor.

"Yes. The superintendent has just told me that there is a possibility that the fire was started by an incendiary. But there is another question — in fact, they are both bound up together. It appears that those cans we saw them taking out contained naphtha, and that the naphtha was there without special permission from the insurance people. But the factory people say the fire started at some distance from the naphtha, and they have the evidence of eye-witnesses that it did start there. Moreover, they rescued every can containing naphtha. The cans were untouched and intact. Yet there will be a controversy, and the factory people, having heard that photographs of the fire were taken, the superintendent thinks they might be first-class corroborative evidence that the fire started on the east side of the wing, the side opposite the storage place of the naphtha, — would head off any trouble with the insurance people."

"Well, well!" was all Allan could say.

"It was the superintendent."

"It did start on the east side," declared Edith. "We all saw it."

"And the camera saw it," added McConnell, with great conviction.

Owen's hands were trembling a little. "I mustn't drop this now," he muttered.

Then there was a rattle at the door below, and a step on the stair.

As the Doctor started forward again a man's head appeared above the stair rail. It was the superintendent. His eyes blinked in an unaccustomed light. For a moment he did not seem to be able to make out the situation.

"I say, Doctor," said the superintendent. "I'd like to be able to say to our president that we have secured these negatives. I'll send you a check for fifty dollars if you'll say they're ours."

"What do you say, Allan?" asked the Doctor, turning about.

Had it not been for the red light Allan probably would have looked very white in the face.

"I suppose they can have them?" said the Doctor when Allan did not seem to find words. "You will be glad that they can be so useful, and — how much did you say, Mr. Superintendent?" the Doctor went on, with an enjoyment of Allan's agreeable stupefaction.

"Fifty dollars," repeated the superintendent. "They'll be worth that to the company. Anyway, I'll risk making that offer. And I want the thing understood. Is it a go?"

"Oh, you can have them!" Allan said. And the superintendent repeated his "Good night!" shuffled his way cautiously down the dark stairs and was gone.

No one said a word until the superintendent closed the door below.

"Fifty dollars!" was all Allan said.

"I hope they come out well," said Owen, fervently.

"I wish I could help," murmured Edith.

"I think that Dr. Owen is doing the best that could be done with the patients," laughed Dr. Hartel.

Then McConnell spoke up. "Gee whiz, Allan, you can buy a folding cartridge camera now!"

IV.

AN ILL-KEPT SECRET.

THE visit of the factory superintendent gave a new excitement to further work on the negatives.

"We certainly shall have to be very careful of them now," said Allan.

"Yes," said McConnell, "they are worth $16.66 apiece — oh, yes, and a fraction!"

"Just to think!" Edith exclaimed, "that this strange thing should happen on the very first night you have your camera."

Owen was holding the third negative between his eyes and the lamp. "This is quite good," he said.

"Yes," Allan remarked. "And you deserve all the credit. We'll have to share the prize money."

"No, no!" Owen answered to this. "You could have done it all without me. It was simple enough."

"It is simple enough when you know how, isn't it?" laughed the Doctor. "Well, Allan, we'll have to leave you with your prize plates. Mind you wash them carefully. If they are to be used as evidence they shouldn't have any questionable spots on them anywhere. To be sure a flaw could be distinguished from a genuine light impression, unless it was a discoloration in the emulsion. That's the reason the negatives themselves are better evidence than the prints from the negatives. The negatives could be tinkered so as to show fire in every window, I suppose. But the negatives themselves would show that this had been done. So many tricks can be done in photography that I don't suppose that a photograph in itself could even be offered as proof. But it is pretty good corroborative evidence when you have the negative."

"That is what this would be, isn't it?" asked Edith, "corroborative evidence?"

"I presume so, if the matter goes so far. They probably can prove by eye-witnesses that the fire started on the east side of the wing, at least I should think so; but the camera will be a first-rate supplementary witness. We can testify that it was absolutely impartial."

"Good fellow," murmured Edith, patting the camera as if it were her dog Sandy.

After the Doctor and Edith had gone Owen and Allan stood the three plates in a bucket and placed the bucket under the tap.

"That's as good a washing-box as any," Owen said, "when you have only three plates."

While the water overflowed from the bucket into the sink the three boys sat on some boxes that were stored in the room, and talked of the fire, and the camera, and picture-making, and Owen related some of the things that happened to him.

"The first pictures I ever made," he said, "were of a railroad train. I took one of the train coming up the track, and one when it was going by. I got them both on the one plate, and it was the worst smash-up you ever saw."

"I suppose every one makes doubles," suggested Allan.

"Oh, yes, they say that even big photographers do it sometimes. And it *is* rough! You see you only make one mistake and lose two pictures."

"It doesn't seem fair, does it?" mused McConnell, rather sleepily. McConnell usually went to bed at nine.

When the plates had been immersed in the flowing water for half an hour, Owen stood them on a near-by shelf, resting them against the wall. "They'll be dry in the morning."

"And then we can make proofs," said Allan.

Under these circumstances it is not surprising that Allan was up early the next morning and out to the dark room to look at his plates. To his great disappointment they were not yet dry. The upper floor of the stable was left without ventilation, and the surfaces of the plates were still moist save for a space of half an inch around the edges.

When Allan consulted his father, Dr. Hartel advised him to open the windows and to place the plates in a current of air. Having done this, Allan started out at once to get the materials for his developing outfit. In view of the fifty dollars that was to come from the plates, Allan thought that he might spend a little more than he first had intended to spend on his dark room. He determined to divide the money with Owen, but even with twenty-five dollars he could, if he chose, buy a fine new camera and still have money left.

On his way to Owen, who was going with him to the Hazenfield stationers' where they sold "amateur photographic outfits," Allan met Cheney, who had seen him at the fire with the camera. Allan did not like Cheney; the truth is, that Cheney did not have a very good reputation in Hazenfield. He had been expelled from the high school, and was what is known as a suspicious and troublesome boy.

"What were you trying to do?" demanded Cheney, "photographing in the dark?"

"The fire wasn't dark," returned Allan.

"You don't mean to say that you tried to photograph the fire!"

"Of course I do."

"Did you get it?"

"Of course I did."

"Well," said Cheney, as if he did not believe the assertion, "I'd like to see 'em."

"They're not mine to show, now," said Allan, with a pride that he could not conceal, even from Cheney. "I sold them."

"Sold 'em!" exclaimed Cheney. "To who?"

"To the factory people. There's something about the factory being set afire, and beside that there's some insurance trouble about where the fire started, and they've given me fifty dollars for the plates."

Cheney gave a long whistle. "Golly!" he cried. "You're right in it, ain't yer! Fifty dollars! I didn't suppose any picture could ever be worth as much as that."

"You see," said Allan, "they can use the pictures as proof if there's any trouble."

"Proof of what?" Cheney demanded incredulously.

"Proof of which side of the wing the fire started."

Cheney smiled as if this idea was very amusing. "I hope you get it," he said, as Allan went on.

Allan felt rather sorry to have said anything to Cheney about the sale of the plates. When he came to think the thing over he could see that the factory people, while they had not said so, might not wish to have the matter known. Allan felt, too, as if he had been boasting, even though Cheney had drawn him into the confession.

He made up his mind not to say a word to any one else, and determined to ask Owen and McConnell not to speak about it. When Owen heard from Allan about his meeting with Cheney and how sorry he was that he had said anything to Cheney, he himself agreed with Allan.

"I shan't mention it," Owen assured Allan. "I haven't mentioned it to any one but mother. She wondered where I had been, of course. I think you had better speak to McConnell."

"I shall," said Allan. "Big McConnell would be just the one to spread the thing everywhere."

"Let us go around to McConnell's now," suggested Owen, "and tell him about it."

They found McConnell sitting in a swing in his yard reading a book. "McConnell is always reading," said Owen.

"Hello!" called Allan. "What are you reading, McConnell? I'll bet it is about an Indian, a detective, or a princess."

"Wrong," replied McConnell. "It's about a farmer's boy."

"And what does he get to be?" Owen asked.

"I don't know yet," McConnell returned. "Maybe a farmer!" he added, laughing.

Allan then spoke about the fire and the pictures.

"I tell you," said McConnell, "I haven't said anything to anybody but the folks at home, and Billy Basset, and the butcher, and — yes, and Mr. Hanford."

"In that case," said Owen, "I guess we might as well tell the rest of the town."

"What do you mean?" McConnell was mystified.

"Only that I thought that maybe we should keep quiet about selling the pictures."

"I see," McConnell assented, "until you get the money."

"Oh, I mean anyway," said Allan. "Perhaps the factory folks may not want everybody to know, and everybody would know if they heard that they had bought the plates."

"That settles it!" exclaimed McConnell, "I'll be quiet. I won't even tell the postman, and I always tell him everything. Have you printed the proofs yet?"

"No; the plates aren't dry yet, somehow. We are going down to get the dark-room things, and we shall get some proof paper."

"I'll go with you," McConnell said, pushing his book under the porch seat.

Little Artie, McConnell's younger brother, came running around the house. "Take my picture taken," he cried to Allan, who was carrying the camera.

"It's funny," laughed McConnell, "but Artie al-

"McConnell was sitting in a swing."

ways gets it that way. He has heard us say 'get your picture taken,' and he always says 'take my picture taken,' or 'take your picture taken.'"

Artie repeated his request, peering into the finder of Allan's box. "All right," said Allan.

"Wait a minute," called McConnell, "wait till I put him on my wheel." Artie was lifted up. "Now," said McConnell.

Artie started when he heard the click of the shutter. "Did you take my picture taken?" he asked.

"Yes," Allan laughed, "I took your picture taken."

"Then let me see it," demanded Artie.

"Not yet, Artie, — to-morrow."

Artie then went back to a box of sand in which he had been playing.

It took Allan nearly an hour to complete his purchases. He had the advantage of Owen's advice, in the choice of trays, for instance. Owen had tried lacquered tin, glass, and pressed paper, but liked rubber best. These cost more, but as he only required three, two to be used in developing and a larger one, capable of holding four plates, for "fixing," Allan felt that he could afford it, — especially with his share of the fifty dollars to come.

Certainly it was great fun to buy these utensils, and the shining glass graduate, — Dr. Hartel had promised Allan a second, large, graduate, as well as a hydrometer and some stopper bottles; the brass-hinged printing frames, the "hypo" and developing chemicals, and the dark-room lamp.

At first Allan had thought of building a dark-room light-box, with red cover glass, in which an ordinary small lamp could be placed; but in view of the

"Artie was lifted up."

probable usefulness of the portable lamp on certain expeditions which he had in mind, he decided to get an ordinary dark-room lamp, and it was a pretty affair. Dr. Hartel had urged Allan to prepare his own developer, at least until he had learned how developers were made up and what properties they had.

All these points had to be talked over, and Mr. Wincher, the stationer, who was an amateur photographer himself, and had on that account come to sell photographic supplies as a department in his store, was patient throughout the selection and offered plenty of advice too.

"Photographers are great fellows [illegible] giving advice," laughed the stationer.

"I'll need quite a supply of advice, too," returned Allan.

Owen and Mr. Wincher did not always agree as to what was best to do. When they agreed on any point Allan was likely to accept their decision. When they did not agree, Allan made the best use he could of their judgment. The truth is that Allan had been reading so much about photography lately that he had made up his mind on a great many points. Dr. Hartel had told him that he must experiment on his own account. "What you will learn from these experiments," he said, "may be worth more to you than the pictures. You must try and learn something from each experiment. This is the only way you will ever really *know* a thing. We act on the best advice we can get to begin with; then we prove to ourselves that what has been said is true or not true — or, maybe, that it is partly true and partly not true. An Englishman once said that he would like photography better if it weren't for the pictures. He was taking a scientific view of the matter. He liked the chemical fun better than the picture fun. I think you will not like the picture fun any the less for taking an interest in the chemistry, and the better your chemistry the more you are likely to find to enjoy in your pictures. But chemistry will give you many disappointments. In any case, you will have disappointments — and they will do you good."

As the boys were all eager to see proofs of the fire negatives, they hurried back to the stable with their bundles, McConnell asking leave to carry the camera.

"There is one thing I like about this sort of printing," said Owen, "and that is that you don't have to have ink. I always get mussed up so with my ink press."

"But when you get the thing started," said McConnell, "you can print quicker on the printing press."

"I believe you can get mussed up in photography, if you want to," suggested Allan.

V.

THE DARK-ROOM MYSTERY.

WHEN the boys reached the coach-house, the plates were found to be quite dry, and after unpacking a printing-frame and slipping from the stiff paper envelope a sheet of printing paper, the first of the fire negatives was soon in the sunlight on a front window-sill.

Allan watched the progress of the printing with excited interest, opening the back of the frame at frequent intervals for a glimpse of the slowly deepening image on the paper.

"That negative wouldn't take long to print," Owen said, "if I hadn't developed it so long trying to bring out everything."

The negative had not looked much like a picture to Allan, and, indeed, the first plate, made as a "snap-shot," showed the strongest lights of the fire scene and very little else. Yet the print gave a meaning to the dark parts of the picture which were blank in the plate.

"It does show the fire; doesn't it?" exclaimed Allan.

"Yes," said Owen, "and it shows that it began on the east side of the wing."

The second plate was much clearer.

"The fire's halfway across in that," remarked McConnell.

The third plate, showing the fire at its worst, revealed even the outlines of the factory. The flames were not so sharply defined as in the quick exposure of the first plate, but the blur made by the yellow tongues of fire was, perhaps, one advantage, and in every other respect the "time" pictures, as Owen called them, were much the better.

Allan unpacked the "toning solution," and with Owen's help toned and fixed the three prints. Owen waited until the prints were getting their last rinsing. "Now," he said, "I guess you are pretty well started, Allan."

"Yes," replied Allan; "and I don't know how I ever should have got along without you, Owen."

"Oh," laughed Owen, "you only would have said mean things about the man who wrote the directions! But you're not through yet! There are more chances for making mistakes in photography than in anything I know of."

After running into the house to show the wet prints (on a piece of blotting paper) to his mother and Edith, Allan set to work on the dark room. McConnell helped for most of the day, whistling loudly while he worked, and telling Allan a story he had read about a pirate who got shipwrecked.

Before supper-time the dark room began to look like a real photographer's den. With an arrangement which he had made for the window, and the strips of cloth around the door, Allan could have the room absolutely dark in the brightest daylight. There already was a long shelf in the room, and an old chest of drawers. After planning a place for everything, Allan made strong resolutions to keep everything in the places he had chosen for them. And he felt much pleased at the way his packages and bottles looked when spread out on the shelf. He scarcely could wait for his father to come in and survey the outfit.

"You have done very well, Allan," said the Doctor, "but you must keep this up; especially, you must keep everything clean, for you will be working in the dark in more senses than one if your bottles and graduates and trays are not clean. Rinse everything after using it, and before putting it away. When your plates act queerly you want to know where the trouble is, — whether in the plates, in the camera, in the time, or in the developer."

Allan said he meant to be very careful, and he put some of his plans into practice when he developed the picture of Artie on the bicycle.

"I tell you, McConnell," said Allan, as his companion was going home, "when you get your camera, we can use this dark room together."

"Well," said McConnell, pleased at being made a partner in so interesting an institution as the dark room, "then I think I ought to chip in some trays and things for myself; don't you think so, Allan?"

"All right!" laughed Allan. "One tray and a plate-lifter."

It was that evening, a little after nine o'clock, when Allan and Edith were studying and discussing the fire pictures, that the factory superintendent came to the door. Mrs. Hartel ushered him into the room.

"Are the plates ready?" he asked. "I couldn't get over any sooner."

"Yes," Allan answered, "they're ready. We were just looking at the prints."

"The prints? Oh, yes!" And the superintendent studied the pictures with great interest. "Great!" he exclaimed, his bushy head bent close to the prints in the light of the centre-table lamp. "Wonderful! They're awfully dark, but you can see plainly where the fire was, and how it worked across. Yes, sir, those pictures may be useful. I'll have the check sent to you to-morrow, my lad."

"I'll go out and get the plates," said Allan.

While Allan was gone the superintendent told Edith and Mrs. Hartel how they had been clearing up the mess at the factory during the day. "Of course," he said, "we have to leave the wing alone until the appraiser comes, and we settle the row with the insurance company. The naphtha cans weren't near the fire; that is, they were on the other side of the partition when it started."

"Do they still think some one set it afire?" asked Edith.

"Well, the factory folks themselves don't think so. We think it started in the packing room, in some rubbish. Fires often start that way. The man they say did it —"

The superintendent got up from his chair when he

"'Think a minute,' said the superintendent."

saw Allan returning. "You've got the plates there, have you?"

"No," answered Allan, his face pale and perplexed. "They're gone!"

"Gone!" exclaimed Mrs. Hartel and Edith together.

"What do you mean?" asked the superintendent.

"I left them standing in a safe place," said Allan, "and they are not there. I have looked everywhere."

"Great Scott!" ejaculated the superintendent. "They couldn't be stolen, could they?"

"Surely not," said Mrs. Hartel. "Are you sure, Allan, that you didn't carry them in here?"

"Think a minute," said the superintendent, with his hand on Allan's shoulder. "Perhaps you put them in some special place. I often do that — and then forget where the place is."

"I know I left them there," Allan persisted, "for I looked at them before I came in to-night. I had them in a place I had arranged for negatives."

The superintendent sat down again. "Have you told anybody about this thing? — I forgot to tell you not to."

Allan declared that he had spoken to but one boy about it, and he enumerated those who knew about it through Owen and McConnell. "The only one I spoke to," said Allan, "was Cheney."

"Cheney!" cried the superintendent. "Sam Cheney's boy?"

"Yes," said Allan, mystified.

The superintendent gave a peculiar grunt. "Do you know," he demanded, drawing his eyebrows together, "that it is Sam Cheney who has been suspected of starting the fire?"

Allan looked amazed and shook his head.

"You don't suppose that Cheney boy could have stolen them, do you?" asked Mrs. Hartel.

"Why not?" the superintendent demanded.

Allan was staring at the lamp. "I believe they are stolen!" he cried. "I remember that I locked that door to-night—and it was unlocked when I went up for those plates just now."

Dr. Hartel appeared at this moment, and the superintendent blurted out, "Doctor, I guess we've got a case for the police here."

"The police?" The Doctor looked his astonishment.

"It looks as if the Cheney boy had stolen those plates."

"Stolen them?" The Doctor listened to Allan's story, and questioned him closely. "It does look like it," the Doctor admitted.

"I guess there's no doubt of it," the superintendent went on. "I guess, though, that we had better not say anything about it just now. I'll quietly have Detective Dobbs put on the case."

"I'm very sorry this has happened," said the Doctor, much annoyed.

"Well, so am I," added the superintendent, "and I don't suppose, Doctor, that you are willing to let the thief go."

The Doctor shook his head. "I'm not willing to let the plates go."

"But we have the prints," interposed Allan.

"Yes," said the Doctor, hopefully.

"I'll go over and see the chief now," said the superintendent, and he went away with a hurried "Good night!"

VI.

DETECTIVE DOBBS.

ALLAN looked dazed when the superintendent had gone. They all looked dazed.

"I am wondering," said Dr. Hartel, "whether Cheney was sent to do this by his father, or whether, knowing that his father was suspected, he did the thing on his own account."

"But we don't know that Cheney did it," said Mrs. Hartel.

"True," the Doctor replied, "but the chances are much that way. Allan," the Doctor continued, "we had better go out and look over the place again."

"A cat might have knocked them down," suggested Mrs. Hartel, as they were leaving.

But there could be no suspicion of a cat. There were no broken fragments anywhere. The only negative in the rack was that of Artie on the wheel.

"And you found the door unlocked?" asked the Doctor.

"Yes," Allan said confidently. "At first I didn't think anything of it. But I remember distinctly now that I locked it, and I remember thinking that it was foolish to bother locking it."

The Doctor shook his head. "It is too bad. I am sorry about Cheney."

Before going back to the house Dr. Hartel made some suggestions as to the keeping of the chemicals, as to guarding the floor from drippings of the hypo, as to pouring from the bottles, as to keeping the place free from dust, and so on.

Father and son were seated talking over photographs and the fire and Cheney, when a sharp rap sounded on the door at the foot of the stairs.

"Come in!" called the Doctor from the top step.

A lank man with a bristling red mustache came up the steps.

"Is this Dr. Hartel?" asked the man.

"Yes," replied the Doctor. "I do seem to be wasting a good deal of my time out here just now."

"The chief sent me over," said the man, "to see you about some pictures that were stolen." As he reached the top step the man looked questioningly at Allan.

"This is my boy," Dr. Hartel said. "He took the pictures — I mean that he *made* them," laughed the Doctor. "You are Dobbs, are you not?"

"Yes," said the man. "I wish you would tell me what you know about it."

"Dr. Hartel made some suggestions."

He was a ruddy-faced man. His mustache stood out like the hairs of a brush, and he had a little red scar over his right eye. When he smiled Allan liked him at once. Allan remembered that he often had seen him down by the railroad station.

They told the detective all they knew and he listened attentively. Then he looked about the rooms, and seemed much interested in everything he saw in the dark room. He held up to the light the negative of Artie on the bicycle, and laughed over it.

"I have a kid about that size," he said. "I wish you'd photograph him sometime."

"I will," said Allan.

"My boy Sporty," said the detective, "is great. Why, sir," said Dobbs to the Doctor, "that kid got a hold of my nippers the other day and got them on the necks of our cat and the cat next door. You never saw such a thing in your life. Scott! wasn't there a row!"

At the thought of the handcuffed cats — that is to say, the neckcuffed cats — the Doctor and Allan joined in the detective's jolly laugh.

Presently the Doctor, wishing to get back to the question of the plates, ventured to ask Dobbs what he thought about the situation.

"Oh," said Dobbs, stooping to pick up a burnt match from the floor, "I guess Cheney did it; though his father didn't have anything more to do with that fire than I had. Say," continued Dobbs, turning to Allan, "how do you light your lamps here?" The room was lighted now by a small, ordinary lamp which Allan had borrowed from the kitchen.

"Why, with a match," replied Allan.

"Will you let me see one of your matches?" asked Dobbs.

Allan took a match from the little tin safe he had tacked up beside the sink, and handed it to Dobbs. As he did so he noticed for the first time that Dobbs had a burnt match in his other hand.

"Then you don't throw your burnt matches on the floor," said Dobbs.

"No," said Allan, perplexed at the statement, "I always put my burnt matches in here;" and the Doctor smiled as Allan indicated another tin box on the corner of the shelf. It was this sort of care of which he had sought to teach Allan the importance.

"I believe you," said Dobbs, "for this isn't your match." Dobbs was holding up the burnt fragment he had picked from the floor.

The Doctor and Allan, coming closer, saw that the two matches certainly did not have the same sort of stem.

"Then that," said the Doctor, pointing to the partly burnt match, "belonged to the thief, whoever he was."

"Looks so," said the detective, briefly, studying the matches, or seeming to. Then "Wait a moment," he said, stepping across the room, and he picked up another fragment of a match. It was almost completely burnt, but the fragment showed that it had been of the same form as the first piece. "He lighted two matches, you see; this one burnt out on him before he found the plates. Then he struck this other one."

Allan's eyes stared. He never should have thought of these things.

"Oh, I wish you had been around to get a picture of those cats with the handcuffs on," said Dobbs, as if that subject was much more interesting. Then he slipped the two fragments of matches into his vest pocket, and when he was going he said: "I don't suppose we'll get down to the fine points on this thing — what you want is to get the plates back if they haven't been broken or thrown into the river. I'll be around again in the morning."

And Dobbs did come around in the morning. "You haven't forgotten about Sporty, have you?" was the first thing he said to Allan. He seemed to have forgotten about the plates, but when he saw Dr. Hartel he remarked that he had been looking into the business.

"What beats me, Doctor," he said, "is why the Cheneys should steal those pictures. If the factory people were right, and the fire did start on the east side of the wing, then Cheney couldn't have anything to do with the fire. The factory people don't think he did. It's the fire marshal who's raising the row. So you see that the pictures help Cheney as much as they help the factory people. If Cheney has stolen or smashed those plates, — I mean the father, — he has removed very good evidence, as I understand it, that he is innocent. I tell you he didn't have anything to do with it. He was sore on the factory management, but he wouldn't be such a fool. That fire started in the east of the wing, nowhere near the naphtha."

"Then why should the boy have taken the plates?" demanded the Doctor.

"I can't see," replied Dobbs, "unless the father or the boy, or both, got it into their heads that these pictures might be used against Cheney in some way. It was a crazy notion."

"Have you any further clews?" asked the Doctor.

"No, I can't say I have. But I threw out a hint

in the Cheney direction that may do some good. The old man is over at Westwall, but I saw the boy. The little rascal actually stumps me. I can't tell whether he did it or not. But I left something there to soak through his thick young skull. It may work."

Dobbs's attention again turned to the camera. "I was saying, Doctor, that I wish your son would take my boy, Sporty." A happy idea seemed to strike Dobbs. "I tell you what I'll do!" he exclaimed. "If you'll make a picture of Sporty, I'll take you to New York with me to-day."

"I'd like to go," Allan admitted.

"I don't see why you shouldn't," said the Doctor.

"Is it a bargain?" asked Dobbs.

"Yes, it's a bargain," laughed Allan.

"You see," said Dobbs, as they walked down toward his house a little later, "I've got to go down to New York anyhow, and you might as well run around with me. I dare say you'd like to see police head-quarters, and some other places, anyway. There are lots of things to photograph down there."

Dobbs lived in a little wooden cottage near the bank of the Hudson. It was painted a bright blue. Allan thought there was something peculiar about the house, and he became fixed in this opinion after seeing more of it. There was a stretch of tree-grown ground back of the house. In the front garden three stalks of corn and four sunflowers were ripening. In the hall-way was a big iron dog, painted blue like the outside of the house — with some of the left-over paint, Allan guessed. In the back parlor were five canaries in cages, all singing in a great clatter of high notes; and a small, but very hoarse-voiced red and green parrot in the corner shouted, "Hello Central!" when he saw

Dobbs and Allan. Mrs. Dobbs, a fat little woman, sat sewing at a window, with a white cat in her lap.

"Where's Sporty?" asked Dobbs.

"I don't know," replied Mrs. Dobbs. "He bothered me to let him paddle, so I put on his trunks and turned him loose."

"Probably he's drowned then," said Dobbs.

But Mrs. Dobbs only laughed softly, and went on with her sewing.

Allan and the detective found Sporty down at a little inlet of the river near the house. He wore red-striped bathing trunks, and was sailing a boat, which he pushed into her proper course with a long stick.

"'Must I put on my Sunday clothes?'"

"Say, Pop," Sporty called out, when he saw his father, "I wish you'd buy me a steamboat." Then Sporty noticed Allan and the camera, and looked curiously at the black box.

"Sporty," said Dobbs, "you're going to have your picture taken."

"Am I?" Sporty peered again at Allan. "Must I put on my Sunday clothes?"

"No, Sporty," said the detective, "no clothes will do. You're just right."

"You mean this way?" Sporty asked.

"Bathing tights are very becoming to your style of beauty, Sporty," the detective went on with his pleasant grin. "Come over here," and Sporty's shining legs timidly carried him to where his father waited. "I want you with your arms folded, Sporty — you know," and Dobbs struck an attitude to show what he meant. "There! that's it!" cried the detective, when Sporty had folded his arms. "Now, don't look so savage. I want you to look as innocent as if the judge was asking you if you had ever been convicted before."

Allan adjusted the camera and pressed the trigger.

"Is it took yet?" asked Sporty.

"Yes," said Allan.

"That was great!" exclaimed Dobbs. But Allan thought Sporty had not looked very happy. "We'll have to take him again some time when he's in the humor," Allan suggested, "or when he isn't looking."

"That's so," Dobbs said; "good idea — when he doesn't know it." He looked after Sporty as the boy went back to his boat, which had drifted far out of her course. "I tell you, he's one of the greatest boys you ever saw. He's simply wonderful. You ought to see him do the cart-wheel. When can I see that picture?"

"Probably I'll develop it to-night and you can see a proof to-morrow."

"Good! Are you ready to go to New York?"

"Could you wait until I changed this plate? I want to take the full plate-holders with me."

"Sure," returned Dobbs; "why not take a lot more. You could change them somewhere — I can fix that."

Acting on this suggestion Allan carried in the pocket of his jacket an extra package of plates when he met Dobbs at the station fifteen minutes later.

"Got all your ammunition?" Dobbs asked. "There's big game in New York; you want to be loaded for bear."

Allan had not been to New York for several months, and now that he had his camera with him the prospect of so many interesting subjects for pictures filled him with a pleasant excitement. It was a bright day, and as he looked out across the glistening Hudson he made up his mind to "do" the Palisades sometime. He remembered a cat-boat cruise he had taken with the McConnell boys, how they almost had been wrecked near Fort Lee. Yes, he thought a cat-boat and a camera would make a good combination. He already found himself planning certain pictures at the base of the cliffs and from the crags overhead.

There were many scenes along the Harlem th.* attracted Allan — the long arch of Washington

"The long arch of Washington Bridge."

Bridge, the varied craft of the river, the loops of the elevated roads; and when they were in Manhattan, there were funny little remnants of the squatter settlement that seemed made — or at least left — to be photographed.

"Now, you understand," said Dobbs, "that I'm in no hurry. You can go anywhere you want to and I'll trot around with you. I want to see how you do it — I'm going to get one of those things myself one of these days."

"You can't learn much from the way *I* do it," said Allan. "I'm only a beginner."

"It seems to me," pursued Dobbs, "that we might do a little of the Bowery and around the Pell Street way — in the Chinese quarter and so on. Oh, I suppose you could put in a week here — slumming with a camera, how would that go? — unless you don't like slums. First, I've got to run in and see one of the Central Office men at headquarters."

And so they went over to Mulberry from the "L" road, and Allan was so much interested in being at police headquarters that he thought no more about the camera until Dobbs was ready to go.

"Suppose we go and look at the Rogue's Gallery," suggested Dobbs, and he led the way into one of the rooms opening off the main hall. "There's a collection of photographs for you!" exclaimed Dobbs, turning the doors of a curious cabinet like a vast wooden book.

Allan stared in amazement at the countless faces that stared out of this curious collection. Something in the style of these faces made Allan feel sad. Yet the faces were not all evil-looking faces by any means. Perhaps it was because they were not that

Allan felt awkward and grieved as he looked at them. There were handsome faces of both men and women, some of them very young — mere boys and girls, sometimes — and the well-dressed and the ragged were shoulder to shoulder. One face, that of a boy who appeared to be of about Allan's age, held Allan's attention until Dobbs asked if he knew the face.

"No," said Allan. He had been thinking how clear-eyed and manly the boy looked. He wondered what trouble the boy had fallen into, and if the boy's mother thought he was guilty.

"Well," said Allan, a lump in his throat, "I don't believe they are all rogues. I believe there were some mistakes — that some of them were innocent."

"Maybe," said Dobbs. "Maybe there are fellows there who don't belong there. And there are a great many folks who belong there who are not there."

Allan said nothing more while they were going downstairs again. They walked back to the Bowery and turned to the south. One of the first things that Allan saw struck him so oddly that he adjusted the focus of his camera to fifteen feet, and turning about made a quick shot, Dobbs watching attentively.

The subject of Allan's picture was a boy of twelve or thirteen perched in a high boot-black's chair with a grimy little Italian "shiner" polishing his shoes.

"Shine?" called the Italian boy over his shoulder, when he saw that some one had stopped; then went on with his work.

"Young America and young Italy," laughed the detective, as they walked down the Bowery. When they came to one of the numerous dime museums of

"'Young America and Young Italy.'"

the street, Dobbs halted and they read over the glaring announcements that plastered the front of the building.

"You haven't a big enough plate to take the fat lady," chuckled Dobbs, "but I don't see but that you might take a shy at the ossified girl. Ha, ha!" the detective laughed loudly, as he pointed to a huge picture spreading across the front of the building; "there you are!" and Allan read: —

> GREAT PIE EATING CONTEST
>
> BY
>
> ELEVEN LOVELY LADIES!!

"Couldn't we get them to let us take that?" asked Dobbs, "when they were half through, you know!"

Allan joined in the detective's laugh, until the man in the little window to the right of the entrance looked over at them with a scowl.

Dobbs appeared to be greatly taken with the idea, and for a few moments Allan feared that he might suggest carrying it out. But presently they left the region of the museums, and Allan changed the subject by catching a group at the drinking tap in front of the Young Men's Institute. Then the detective and Allan came to the shooting-galleries where wooden deer and green lions were ceaselessly jumping, and silvery balls were rising and falling on jets of water until shattered by some successful marksman.

"The drinking tap."

"Here's a chance," said Dobbs, "to get some of those queer things in animal locomotion. The great advantage here, though, would be that no matter when you caught the deer and lions, their legs would always look perfectly natural. That would be a big advantage. I don't like these snap pictures that show the horses standing on one foot with their hind legs twisted."

Below Grand street, in front of a clothing store, they found five men standing at the curb, each supporting a huge wooden letter. When Allan stepped into the street to read the letters from the front, he found that they spelled "P–A–N–T–S." A "Great Pants Sale" was in progress here. Other sales were in progress at every step of their walk.

"A Great Pants Sale."

"Isn't it almost time to eat?" asked Dobbs. "I know a joint here where you can get the best steak in New York."

The "joint" was across the street, and Dobbs saluted the proprietor cheerily as they walked in. Allan was too much stirred up by the sights and sounds out of doors and the whirr and clatter of this "quick lunch" indoors to eat much. When they had finished Allan wished to share the bill.

"But there isn't any bill," said Dobbs; which seemed to be true, for, as they passed out, Dobbs merely nodded again to the proprietor. "I once did that fellow a great big favor," said Dobbs, "and so my money wouldn't go in there — nor yours either if you were with me."

"Suppose you lived near by," said Allan, "and

took all your meals there, do you think he would keep on being willing?"

Dobbs laughed. "Well, I think I could keep it up quite a long time, if I wanted to."

Presently they came to Chatham Square, and Dobbs, pointing over to the west, said, "Here's Chinatown."

Dobbs knew all the queer places, the joss houses, and theatres, and restaurants, and he pointed these out or indicated them with a jerk of his head.

"They are suspicious here," said Dobbs, "and when you want to take anything I'll turn and talk to you so that you can shoot past me." It was in this way that Allan caught a group of men in front of one of the queer places. Hundreds of Chinamen were passing up and down the street, hundreds more could be seen in the tea-shops and at the windows of the restaurants.

Suddenly Dobbs turned away with an exclamation which Allan did not understand, and, looking about, the boy saw the detective run into an alley.

Allan, much perplexed, walked to the mouth of the alley and peered into its dingy depths. But there was no sign of Dobbs. He had utterly disappeared.

VII.

IN NEW YORK.

ALLAN could scarcely believe his eyes. The detective certainly ran into the alley, and there seemed to be no escape from the alley save by the flight of wooden steps running up the side of the building on the left. These steps were in full view to the top, and it certainly was impossible that Dobbs should have mounted these in the brief moments that had elapsed before Allan reached the point from which he could see them.

For several minutes Allan stood there in perplexity. Then he walked as far as the corner and back again to the alley. Still no sign of Dobbs. A Chinaman in a tea-shop came to the door and stared at Allan. Other Chinamen across the way seemed to be wondering what he wanted there. Two barefooted Italian boys stopped, and very deliberately examined the camera and Allan.

Allan again walked to the corner, and as he turned he saw a little ragged boy enter the alley, and he lost no time in retracing his steps. But when he reached the alley the boy was not to be seen. He, too, had melted away.

Allan then determined to settle this mystery if he did nothing else, and to wait there until another figure attempted to elude him. "I'll follow them," he said to himself.

The Chinaman in the tea-shop saw Allan take his stand at the curb opposite the alley. For five minutes or more he stood there watching the life of the street, peering at the strange signs and banners and balconies. He looked toward Chatham Square, from which came the clatter of the elevated trains and the roar of street traffic. Occasionally he turned to the alley as if expecting to see Dobbs reappear from its shadowy depths.

Once, when his eyes turned to the alley, he saw a man with a red shirt shambling toward him, a lame man who leaned against the sides of the alley to support himself. But it was impossible to see where he had come from. The mystery was as deep as ever — as deep as the alley.

Presently a woman who passed him with a bundle of clothes entered the alley with a rolling step, and Allan instantly followed her, to the evident perplexity of the Chinaman in the tea-shop. He followed so closely at the woman's heels in his determination not to let her melt out of his sight, that the woman glanced back at him over her shoulder in suspicious inquiry.

Then Allan discovered what many another discovered long ago, for when the woman reached what had appeared to be the end of the alley, she turned to the right, and the boy, following, found himself in another alley running at right angles to the first.

It seemed very absurd that he had not thought of this before. Several doors opened into this second alley, and through one of these the woman passed with her bundle.

Allan went straight ahead into the cross street upon which the alley opened. If Dobbs had been playing a trick upon him, it might be that he was waiting here somewhere. But there was no sign of the detective.

Whether it was a joke or not, Allan made up his mind not to worry any more over finding Dobbs. Possibly the detective might come back to the place from which he had disappeared. Perhaps he was in one of these houses. But there was no way of telling where he was, or when he would come if he did come. Allan had been in New York before. He knew his way to the Grand Central Station. That did not trouble him. Yet he was disappointed in losing so good a guide as Dobbs.

However, there were still two hours of picture-making daylight left, and he determined to use these as best he might on his own account. They had not lingered in Chatham Square, and Allan walked again in that direction, and down into Chatham Street to Newspaper Row and the City Hall. On his way he came across a pretzel man with a store of salty pretzels strung on a stick.

In City Hall Park Allan found a group that excited his interest. The group of men and boys were in a circle, and made up such a good picture that Allan sighted his camera at a distance of twenty feet and pressed the trigger. Then he went forward to satisfy his curiosity.

"The group of men and boys in a circle."

"What's the matter?" Allan asked of a man who was turning away.

"Craps," answered the man, grinning.

Pressing forward into the group, Allan heard the click of pennies and caught a glimpse of a boy's grimy hand tossing some dice on the stones. When some one uttered a peculiar exclamation, the owner of the grimy hand and half a dozen other boys darted out of the centre of the crowd, and fled in great haste, leaving on the stones one of the dice, which an on-looker picked up. Then Allan understood that the boys had been gambling and that some one had sighted a policeman; a fact which gave him food for thought as he crossed the square again and walked up Centre Street. He had determined to move again in the direction of the Grand Central.

While he was thinking of the boys and the policeman and how much of life in this part of New York seemed to be made up of a battle between the two, he caught sight of the Tombs prison, which had been half torn down and patched up again since he saw it once before.

It was while he was picturing the Tombs from the corner of Franklin Street that a boy who had watched everything he did from the moment he stopped there, tapped him on the sleeve, and said, "Say, d'yer want a good picture?" The boy tossed his head in the direction of Baxter Street with a wink so jolly that Allan concluded that the suggested subject must be amusing at least, but he followed the boy for the space of half a block, when the boy, who had been trotting ahead, halted with a laugh before the steps of a dirty, empty-looking house.

On the steps was what appeared from a little distance like a bundle of soiled rags; but when Allan drew near he saw that there was a living creature in the rags, — an old woman lying as if she had fallen there, a rumpled black bonnet in her lap, her head resting against the rail, and her yellow, wrinkled face upturned to the sun.

The boy giggled as if he thought it all a vast joke; but Allan shuddered and looked about as if in wonder that no one had come to help the woman; and when he saw a fat policeman strolling toward him he hurried forward to say: "Officer, here's a sick woman. Shouldn't some one get her out of the sun? She may be dying."

The policeman looked at Allan with an expression which Allan did not understand. For a moment the policeman looked at the woman; then he spat into the street, and said, "I don't think she's dyin' yet," smiled at Allan, and continued his walk.

Allan's face grew hot, and he wanted to shout, "You're a brute!" after the policeman, when a girl came out of the doorway of the house where the woman lay, and seeing the object on the steps, came forward and began shaking the woman as if to arouse her. The girl had a sad face. Allan thought she looked as if she had cried very often.

The woman opened her eyes finally, and Allan, placing his camera on the steps, helped the woman to rise, and by the aid of Allan and the girl the woman tottered up the steps and through the doorway.

"We live upstairs," said the sad-faced girl, quietly. Allan knew that this was a request to help a little longer. It was hard work on the stairs, for the steps were steep and narrow, and the old woman trembled violently.

When they had reached the top, the girl, with a grateful look, said, "I'm much obliged." The old woman did not speak. As he came downstairs an ugly girl with a baby in her arms said to Allan, "Mrs Grimmins is drinkin' very hard again."

Allan went out without a word. He was so much upset that he did not notice at first that his camera had gone. Almost at the moment when he did discover his loss, he saw the camera in the hands of a boy who was scudding around the corner.

Allan was the best runner in the Hazenfield high school nine, and a hundred feet beyond the corner he came upon a group of boys who had the camera be-

"He caught sight of the Tombs prison."

tween them, and who in another moment would have been out of sight in one of the alleys.

"I'll take that," said Allan to a big, rough fellow who had his fingers on the carrying strap.

"Who said so?" was the response.

Allan caught hold of the camera, but the big fellow held fast, and gave Allan a violent push with his left hand. A little crowd sprang up around Allan instantly, and several of the boys began to jostle him and to pull at his coat.

Allan knew that in trouble of this sort it was necessary to get rid of the biggest enemy first, and, still holding the camera with his left hand, he struck his biggest enemy squarely in the face. As the other fell sprawling over the sill of a grocer's shop, Allan wrested the camera free, and, turning about, he struck quickly at two of his other assailants, clearing a space about him.

It was a very uneven affair, for Allan was hampered by the camera. Each of the others had two hands to his one. But Allan fought furiously, and might have made a very good defence with his single hand had not his big enemy, regaining his feet, approached Allan from behind, and, throwing his arms about him, flung him to the walk.

"Soak him, Pete!" yelled several of the boys, gathering for a chance to use their feet.

Pete was powerful, but in agility he was no match for Allan. In a moment Allan had Pete under him; but he had lost hold of the camera.

One of the boys, with a shout, grasped the black box and started to run. He did run — plump into the hands of a policeman.

"The cop!"

The crowd melted in a second's time. Only Allan, Pete, and the boy who had the camera, remained to give an account of the affair.

For a moment Allan thought it was the same policeman he had seen on the other street. But it was not. This new policeman held fast to the boy who had the camera, and who seemed to be wishing that he didn't have it.

"They have been trying to steal my camera!" cried Allan, adjusting his crushed hat.

"He lies!" roared Pete.

"Well," said the policeman, lazily, and as if there was nothing exciting about the incident, "I'll take you all in till we talk it over, hey?"

The boy with the camera began to whimper. "I was only takin' care of it!" he cried.

"I never touched it!" protested Pete. "He stole it hisself!"

"Let's take a walk," said the policeman, holding the small boy and Pete each by the collar. "You walk ahead," he said to Allan.

"Do you mean that you are going to arrest *me?*" demanded Allan.

"I mean that you're goin' to walk," returned the policeman, using Pete to push Allan forward; and the policeman with his three prisoners moved toward Elizabeth Street.

VIII.

TWO ARRESTS.

ALLAN, angry and chagrined, did as he was told and led the line; and very soon a crowd began to gather again. Even some of the boys who had most wisely taken to their heels, seeing that the policeman had his hands full, ventured again into close quarters, and made audible remarks to Pete and his companion.

They scarcely had gone a block when at least a hundred ragged boys and girls of the quarter were following behind or running ahead to shout the news into the alleys. Recruits came running from Mulberry Square. A man with a push cart drew near to the curb and kept abreast of the procession. It seemed to Allan that at least a thousand women were craning their necks out of windows or crowding eagerly in doorways. His face reddened, and he did not know where to look. He certainly felt very angry and resentful, and yet, when people peered at him, he wondered whether they did not think he looked guilty of something.

"What did they do?"

"They were caught lifting."

"Three of them."

"They tapped a till."

"They got the whole gang."

"There was a fight."

These and a hundred other comments and inquiries came to Allan as he threaded the crowded sidewalk; the shuffle of the policeman, with Pete and his companion behind, and a clatter and patter of a multitude of feet everywhere.

Allan was wondering whether they were going to a patrol box or directly to some station, and they were within a short distance of a cross street, when there were signs of a new and separate commotion ahead. Heads in the windows began to turn the other way, and the advance guard of the procession in which Allan had been moving had darted further ahead, and was mingling at the corner with another crowd that surged toward the same cross street.

Then it became plain that there had been another arrest somewhere. Allan could see the helmet of a policeman, the face of a man without his hat; and as they swung around the corner, the two processions almost at the same moment, Allan saw another face that looked familiar.

Yes, it was Dobbs, — Dobbs holding one arm of the man without a hat, — Dobbs, rather flushed, Allan thought, as if he had been running, or was excited. Dobbs looked across at the other crowd and saw Allan, for he laughed and waved his hand, and his lips moved, but in the hubbub Allan could not hear what he said.

The police station was in Elizabeth Street. It looked very gloomy, somehow. Dobbs was laughing again and waving his free hand, just before he turned into the doorway of the station. The crowd had grown to great proportions by this time. Allan never had suspected that so many men and women, who looked as if they might have something better to do, would take so much interest in seeing four prisoners taken into a police station. Several policemen on the sidewalk seemed to be amused; indeed, the affair appeared to be something of a joke to many people, including Pete.

"Several policemen on the sidewalk seemed to be amused."

Allan, followed by the policeman and the two other prisoners, found his way through the police station door.

When they got in Dobbs was talking over the railing to the Sergeant at the desk. When he turned and saw Allan, he called out a cheery "Hello! Did you wonder where I went to? So you've been snapping an arrest, have you?"

"No," said Allan, solemnly. "I've been getting arrested."

"You?" Dobbs laughed in a puzzled way, and looked at the policeman, who was pushing Pete a little further away from the door. "Arrested! Wait a minute," and Dobbs turned again to the desk, while the Sergeant wrote something in a book, and the man without a hat answered certain questions in a low voice.

Although he was feeling decidedly uncomfortable, in spite of finding Dobbs again, and although there was a great chorus of voices in the street and a crowd of faces at the door, Allan found himself watching the face of the man without a hat. It seemed to him that he never before had seen a face so white. Once the man turned and looked at those who were standing near him. He had extremely dark eyes, that twitched — sad-looking eyes, Allan thought.

When the man was led away toward the back room Dobbs swung about quickly, and said, "What's this? Arrested? What for?"

"They stole my camera," Allan replied, "and I was trying to get it again."

"How about this, Steve?" asked Dobbs of the policeman who had captured Allan, and who was now leaning lazily against the railing.

"I dunno," returned Steve. "They were in a mix-up when I got there. This young rat here had the camera," and he pointed to the smallest boy. "Do you know him?" he added, pointing to Allan.

"Why, he's my neighbor!" said Dobbs, who evidently was much amused. "This is his camera. I've been taken with that myself. I just left him half an hour ago, when I first spotted the ghost."

"I thought it was his," pursued the policeman, though Allan looked savage, and didn't believe him. "Thought I'd bring in the whole debating club, and let them have out the scrap here. I suppose Pete was trying to win that box."

"Well, we'll put the pair in for this," said Dobbs, frowning at Pete and the other boy. "Did they hurt you?" he asked Allan.

"No," Allan answered, "but I think I hurt them a little."

Pete's upper lip was swollen until he presented a comical appearance. Dobbs saw this, and a twinkle came into his eyes. "You young highwayman!" he growled at Pete.

"I was only foolin'," whined Pete.

"Well," demanded the deep baritone voice of the Sergeant, behind the desk.

"I wish you'd let them go," protested Allan. Dobbs was picking up the camera.

"What!" growled Dobbs, with something in his voice that made Allan understand that he didn't mean it; "let these bandits go?"

"We didn't know whose it was," whimpered the smallest boy.

"Of course," snorted Dobbs again, "you were looking for the owner, weren't you, like a good little boy?"

"No complaint?" asked the Sergeant, in a dry tone.

"Now, Sergeant," said Dobbs, holding up the camera and blinking into the finder, "please look pleasant; it may hurt your face to do it, but look sweet for just a moment."

"What about this, Steve?" demanded the Sergeant, turning to the policeman.

"He won't make a complaint," said Steve.

"Then get out of here!" ordered the Sergeant, in a terrifying voice to Pete and his companion, and those two reprobates did get out with wonderful agility.

After they had gone Allan was surprised to notice what a pleasant smile the Sergeant had. Dobbs went on to tell Allan, and the Sergeant at the same time, how he had caught a glimpse of the man with the white face, whom he called the Ghost; how the man darted into the alley; how he had pursued him through the two alleys into the side street, and into other alleys; how he had lost him, summoned the assistance of a policeman, and searched several houses for him; how they caught him at last stretched on his face under the rafters.

"And I have been looking for him for three years!" chuckled Dobbs.

"What did he do?" asked Allan.

"Do? What didn't he do? That fellow's been bad ever since he began to breathe. We want him in Hazenfield for a store robbery and nearly killing a watchman. Did I worry you some by running away?"

"I *was* bothered a little," admitted Allan.

"Well, I've got to leave the Ghost here until court hours to-morrow, and I'll be going back to Hazenfield soon. Suppose we take another turn around before we go back."

"I'll have to change my plates, somehow," said Allan.

"All right," and the detective went back to find the doorman. Presently he returned with a sprightly, gray-haired man at his elbow. "The doorman says he has just the place for you here."

The place suggested by the doorman proved to be quite what Allan needed, as far as being dark, for when he had closed the door there was not a speck of light anywhere. It was a large closet with an old trunk in one corner, an old coat with brass buttons hanging over it. A musty smell pervaded the place, a rat scampered somewhere in the darkness, and Allan did not especially enjoy the interval during which he transferred his used plates to the box, and the new plates to the holders.

Dobbs was outside guarding the door, though the doorman said that no one ever would think of opening it.

"Ah, Captain!" cried Dobbs, as Allan emerged from the closet. "Loaded for bear now, are you?"

The doorman took great interest in the camera, and so did the Sergeant. Allan felt very grateful and generous, and suggested taking the Sergeant at his desk, the doorman standing near with his keys in a military attitude.

"I guess I should count six," said Allan, "and you mustn't move."

"Hear that, Sergeant?" demanded Dobbs. "Look benevolent and don't breathe."

At this the Sergeant's lips twitched, but he held quite still until Allan had completed the exposure.

The thing was hardly done when a policeman came in with another prisoner.

"Business is good to-day," laughed Dobbs; "good-by, Sergeant; hold on tight to my Ghost."

Then Allan confessed to Dobbs that he had forgotten to change the focus of his camera to short range, and that the picture certainly was spoiled.

"Let it go," Dobbs said reassuringly. "They'll forget all about it, anyway. What do you say to walking up to the Grand Central by way of Broadway? We can take in Union and Madison Squares and so on."

Allan thought this a good idea. The truth is, he was feeling resentful toward the regions in which he had spent most of the time since reaching the city in the morning. He was glad to get away toward Union Square and up-town.

Dobbs made many suggestions as to pictures, but Allan did not find it to be possible to act upon many of them. He made some pictures in the squares,—a tramp asleep, the Plaza at Twenty-third Street, carriage-crowded Fifth Avenue, the kaleidoscopic bustle at the Grand Central.

When they were on the train again, Allan began to feel that it had been a notable day. Taking the pictures had made the trip seem more interesting than ever before, and the arrest —

He hated to think about the arrest. It had happened by no fault of his, unless it might be the fault of leaving the camera where he did when he helped carry the old woman into the house. But he felt soiled by it, and grew red in the face again at the thought of the crowds that had looked at him and perhaps measured him as a criminal.

"A tramp asleep."

When they parted Dobbs made Allan promise to let him have a proof of Sporty's picture as soon as he could. "I'd like to have a squint at those others too," he said.

Allan himself was eager to see the results of his day's work with the camera; and although his mother and Edith were much absorbed in his account of the day's incidents, he spent the half-hour before supper in preparing the dark room for the developing.

Edith followed him and had a score of inquiries as to the Bowery, and the mysterious alley, and Broadway, and the battle with the boys.

"And just to think, Allan, those horrid police

might have locked you in a cell like a common criminal."

"Anyway, I was glad to see Mr. Dobbs," said Allan. "He used to be on the force in New York, and he knows everybody."

"And I hope, Allan, you'll never want to go in those dreadful places any more. I like pictures of pleasant places and nice-looking people."

"It takes all sorts of people to make a world," said Allan, rinsing a tray under the tap.

"Yes," admitted Edith, "but you don't need to mix with all sorts."

After supper Allan began his developing, and Dr. Hartel was with him for fully an hour, long enough to see the picture of Sporty and the Bowery shoe-black, and the group of boys around the "Lemonade Man."

It was all very fascinating, this work in the red glow of the lamp, — the moments of expectation until the first signs of the image appeared, the slow growth of the picture under the ripple of the developer, the glimpse of the clear negative after the fixing. The trickle of the water from the washing-box was real music to Allan.

Of course there were disappointments in some of the plates, resulting from mistakes in the focussing, from intrusive foreground figures, from too rapid movements that made a blur. But there were compensations too, for there were many unexpectedly interesting things in the pictures, things not seen by Allan at the moment of pressing his trigger, funny gestures of people, droll expressions of faces.

It was nearly ten o'clock when Allan left his last plates washing and went into the house to report on his successes and failures. He carried with him a

"The 'Lemonade Man.'"

rack holding the first dozen of the plates, which he wanted to study in the better light of the sitting-room lamp. This lamp had a plain ground-glass shade, which made just the right relief for the image of a negative.

Half an hour later, when Allan returned to the stable, he found the door at the foot of the stairs slightly ajar. This reminded him of the night his plates were stolen. It set him thinking very quickly.

He had closed the door when he left the stable. He remembered turning about with the rack in one hand while he drew the knob with the other.

Yes, he was sure he had closed the door. He stepped within the doorway, and almost as he did so he heard a step on the stair and a shuffle as if some one were crowding against the wall in the shadow.

"Who's there?" asked Allan.

There was no answer.

"You might as well speak," continued Allan; "you can't get away."

No sound came in reply to this.

Allan opened the door as far as it would go, and as he did so a figure arose in front of him and roughly tried to slip past him. Allan was too quick for the figure. He caught it with both hands — for happily he had not carried anything with him to the stable — and with all the force at his command threw it back against the steps.

The figure grunted at this but gave no other sign that might help to its identification.

"Who are you?" demanded Allan again, panting with his exertions to hold the wriggling unknown, who presently worked his way off the steps and with a quick leap to his feet had almost reached the door, when Allan caught him again and the two dropped in a heap across the sill.

The light from the house now fell on the face of the unknown. It was Cheney.

IX.

GREAT EXPECTATIONS.

CHENEY was very much frightened, but he cried sullenly, "You let me go!"

Allan did let him up from the ground, but still held fast to him. "So you *are* the thief, Cheney? And you want more, do you?" Allan's voice trembled. "Cheney, I'm going to hand you over to the police. You deserve it. You robbed me, and now you were trying to do it again."

"No, I wasn't," whimpered Cheney. "Let me go, Allan! Don't have me arrested. Please, Allan!"

Then Cheney suddenly gave a violent twist of his body, hoping to catch Allan unawares, and escape. But Allan's fingers never loosened their hold.

"It's no use, Cheney," Allan said, speaking as

quietly as his excitement would permit. "It wouldn't do you any good to get away, anyhow. I know who it is, and you would be arrested before morning."

Cheney began to cry. "Don't, Allan."

"But you are a thief, Cheney."

"No, I ain't, Allan; I did take them, but I just put them back."

"Put them back?"

"Yes, Allan, they're up there all right."

"Do you mean that you *have* brought them back?"

"Sure," Cheney answered fervently.

"Well, let us see," said Allan. "You go up first;" and he made Cheney go ahead of him up the stairs in the dark — for he had absently blown out his light when he went into the house. Allan struck a match at the top step.

"See! there they are!" and Cheney pointed to the three plates lying on a chair, the first object Cheney had encountered in the dark.

Allan picked up the plates, holding the match in his other hand. The match burned out, and he struck another, and lighted his "white lamp." Then he looked at the plates, one by one, to see that they were not injured. Convinced that they were not harmed, he turned to Cheney, who stood falteringly and uneasily watching him.

"Well, you can go, Cheney, for all I care."

Cheney immediately began to recover his self-possession.

"Oh, I was only foolin', Allan. I was goin' to bring them back."

"You *did* bring them back," said Allan. "That saved you. I guess I can get Detective Dobbs to let you go."

"Detective Dobbs," stammered Cheney; "you're bluffing."

"Well, Dobbs isn't," said Allan. "He suspected you from the first. He spoke to you about the plates, didn't he?"

"Yes," Cheney admitted, "but I didn't think he knew anything. He only said some pictures were stolen, and if I saw anybody with them I better tell him to get them back in a hurry."

"Look here, Cheney," demanded Allan, stepping close to the other, "what did you take them for?"

Cheney was staring at the floor. Then he lifted his head. "Oh, I told you," he said. "Just for fun."

"No, you didn't, Cheney."

But Cheney would confess nothing further; and when it occurred to Allan that asking Cheney why he took the plates was tempting him to confess that he had thought to shield his father, he decided to say nothing further about the matter to Cheney. The plates were there again — that was pleasanter to think of than proving any one to be a thief.

"Let it go," was all that Allan finally said to Cheney.

"You're not going to say anything about it, are you?" asked Cheney.

"No."

And then Cheney shuffled down the steps, and was gone.

Allan intended to take his newly developed plates out of the washing-box and place them in the rack. He scarcely took time to glance at them. The important thing at that moment seemed to be that the fire negatives were safe after all.

When he ran into the house, Allan held the three

plates aloft and cried out to his mother and Edith, "Guess!"

"Something good you caught to-day?" asked Edith.

Mrs. Hartel saw something different in Allan's face. "The fire negatives!" she said,—"you found them!"

"They *are* the fire negatives," Allan said exultantly, "but I didn't find them exactly,—the thief brought them back."

"The thief—brought them back!" exclaimed Edith.

Allan nodded. "And they're not scratched. Isn't it lucky?"

"Who was it?" asked Mrs. Hartel.

"Cheney."

Mrs. Hartel shook her head regretfully. "I'm very sorry it was Cheney."

Allan related the incident of the meeting, the struggle, and the confession.

"I think you acted rightly," said Mrs. Hartel. "It will be as well to say nothing more about it."

"Do you really think," asked Edith, "that he took the plates on his father's account—because he was afraid the pictures might prove something against his father?"

"I don't know," said Allan. "He wouldn't confess—and I didn't much care to make him. Anyway, I'm quite sure his father had nothing to do with the fire. Mr. Dobbs doesn't think it was set afire, and the factory people don't either."

"I'm glad," said Edith. "It seems dreadful to think of—any one deliberately setting fire to a building."

Dr. Hartel did not reach home until late that night, but Allan waited up for him. He wanted to tell the news himself.

The next morning he hurried over to Owen's and to Detective Dobbs's house. Dobbs only remarked, "I thought so." The detective seemed to take it for granted that Cheney would go unpunished. "But he should be thrashed for it," he said, and added: "How about Sporty's picture?"

"I think it is going to be good," Allan replied. "I'll fetch you a proof to-day." He could see that Dobbs was more interested in Sporty's picture than in anything else, though he seemed to enjoy all of the proofs of the pictures made in New York which Allan had to show him that afternoon.

Meanwhile, Allan had gone to the factory superintendent and delivered the plates to him. The superintendent was delighted.

"I don't know whether they are going to be of any use to us," he said, "but I hated to lose them that way, and they may be very important. Matling!" called the superintendent to a man at a desk in the corner of the factory office, "make out a check to young Hartel for fifty dollars, and take it in to Mr. Ames."

Allan gave his full name; the check was signed by a white-haired man in the adjoining room, who came out from the inner office as the cashier was handing the check to Allan.

"So you are the boy who photographed the fire," said the white-haired man.

"I guess he would like to do it often," laughed the superintendent.

"Probably — if it paid as well. But it is no more

than the plates are worth to us just now. I want to see them." And the white-haired man examined the plates with great interest. "They don't look much like the prints, do they? I tell you," he said, turning again to Allan, "you must let us call on you sometime when we want a piece of photographic work done. We probably shan't have a lawsuit in mind again, but there are things we need from time to time, and you could make use of a little money now and then, I suppose — photography costs something, doesn't it?"

Allan said that he should be glad to try his hand at anything that he was able to do. "You see, I'm only a beginner," he said. He liked the white-haired man, who had a pleasant look in his eyes and whose smile was very friendly.

When Allan left the factory the feeling of good fortune, of having succeeded after threatened failure, of new opportunities for his photographic enthusiasm, not to mention the feeling of the check in his pocket and the new privileges it promised, gave him a cheerful expression of countenance which doubtless appeared to Owen when he met him on his way back to the house.

"What luck?" asked Owen, cheerily.

"Good luck," answered Allan. "Everything is all settled. They paid me for the plates, and the old gentleman — the president of the company I guess he is — thinks there are other things I might do for them."

"That's great!" exclaimed Owen.

"And, Owen," Allan went on, "I shan't have a bit of fun out of this money until you have divided it with me."

"No, no!" exclaimed Owen, sincerely, "I don't

"'Well,' interposed Owen. 'I tell you what we might do.'"

think I should have any of that. I only helped you; it was your camera."

"But, Owen," insisted Allan, "you did a good deal more than help me; you really did most of the work. Anyway, I couldn't enjoy the money unless you shared it. It wouldn't seem fair."

"But I couldn't feel comfortable, either, if I took it."

"Why not, Owen? You could use it getting some new stuff, and —"

"Well," interposed Owen. "I tell you what we might do. Ever since I saw those rooms in your coach-house, I have been thinking that it would be a fine idea for us to have them for a club — a camera club, if your father would let us have them. Now, if you really think you couldn't be happy with all that money, why not take some of it and spend it fixing up the rooms and getting things for a club?"

"Splendid!" cried Allan. "It wouldn't seem so good as giving to you, Owen; but it would be great to have a club, and we could all have the use of better materials than we could afford to have on our own account."

"Besides," continued Owen, "McConnell was with us, and he would feel badly if he wasn't counted in. He told me to-day his brother was going to give him money for a camera next Saturday, and it would be right to count him in as a — what do you call it? — charter member of the club. After that the others who came in would have to pay an initiation fee."

"Yes," Allan assented, "we must have McConnell."

"Do you think your father would let us use the coach-house?"

"I'm sure he would," Allan declared. "There are

three rooms there he never uses, and — hello! there's McConnell now."

McConnell was on his wheel, and was riding with his hands in his pockets — a trick to which he was addicted. When he saw the boys he made so sudden a movement to extricate his hands and grasp the handle bars for a quick stop, that he had a narrow escape from a tumble in the gutter.

"McConnell," said Owen, "if you think you can do that again, I'll go and get my camera."

"I hope you didn't worry," said McConnell, coolly. "I can stop and dismount with my hands in my pockets and not spill the machine either."

"You ought to have joined the circus," Allan said.

"Or Buffalo Bill," added Owen.

"Oh, say!" exclaimed McConnell, "did you hear that Buffalo Bill was going to be at Fitchville next month?"

"The Camera Club will have to go over," said Owen.

"The Camera Club?" queried McConnell.

"Yes, McConnell; you didn't know, did you, that you are a member of the Camera Club?"

"Am I? Without a camera?"

"Well, you'll have one before the club gets ready to be a club."

"I will have one soon," said McConnell. "I hope next week. But what do you mean by the club?"

"We have been talking it over," Allan said, "and Owen thinks we might have a club in our coach-house rooms. I got my money from the factory people to-day."

"You did?" exclaimed McConnell; "without the plates?"

"But I got the plates again;" and Allan explained the situation to McConnell, how they had decided to use half the money in fitting up the rooms, and that Owen, McConnell, and Allan were to be charter members of the club.

McConnell did not conceal his delight over this news.

"Then I must have a camera!" he cried. "And are you going to get a new camera, Allan?"

"Yes, I think I'll get a cartridge Kodak."

"Then why not sell me your Wizard?"

"I want to sell it, and if you'd like it — "

"I would like it," said McConnell. "I know what it can do."

Allan mused a moment. "But I haven't thought anything about what I should sell it for. It's second-hand now."

"Oh, it isn't *very* second-hand," McConnell said.

"It is a fifteen-dollar camera," said Allan. "Would ten dollars be too much?"

"I think that would be very fair," said Owen.

"So do I," added McConnell, "and that's just what Bill has promised me."

"Then we are all fixed!" laughed Owen. "Suppose the club goes and has a meeting?"

"I'll agree to that," said Allan, "if you fellows will go with me to Wincher's while I get my cartridge camera. I can't wait."

"That's all right," said Owen, "we'll have the first meeting of the club at Wincher's; and besides, we want to see what he has that we can use for the club, don't we?"

"Surely," Allan assented, "and the club will be on hand to say what it wants."

McConnell was immensely happy, and turned a handspring before getting on his wheel.

Allan knew that Wincher had a "four by five" cartridge Kodak in stock, and he went straight to the point as soon as he arrived at the store. Wincher put out the camera on the case, and Allan lifted it with affectionate eagerness. They opened the folding front, and Wincher reached for a film cartridge and showed Allan how it was placed in position by removing the sliding back of the box. It was all very pretty and ingenious and simple, and the boys were delighted. To make Allan's satisfaction complete Wincher gave a cartridge with the camera for the twenty-five dollars.

"And now," said Allan, "the club can do as it likes with the other twenty-five."

"Don't you think," said Owen, "that we had better speak to your father first, and then make our plans?"

"If you like," Allan said; "but I know it will be all right. And then we might look around the rooms and decide what we need."

Dr. Hartel was much pleased with the plans for a club. He promised the boys several chairs, a box-scales, and other furnishings, which he had in mind, and gave them some advice as to rules which they must adopt as to the use and care of the dark room. It was agreed that the rules should be written and posted in the rooms.

"You see," said the Doctor, "when your new members come in there would be confusion if you did not have working rules and regulations."

The boys had not thought much about the other members. "[illegible] never would be any more," said McConnell, "[illegible] ...as."

"But your new members will bring more funds and you can improve your outfit. However," added the Doctor, "you must limit your membership."

"Do you mean, The Hazenfield Camera Club Limited?" asked Allan.

"No," laughed the Doctor, "that Limited means something different. I mean that you must decide now to have a certain number of members at the most — say, twenty-five."

"Twenty-five!" exclaimed Allan. "Where could we put them?"

"Oh, they wouldn't all be there at the same time. But you boys must plan all these things for yourselves. Find out how the best camera clubs are managed, and follow their example. Make yourselves at home in the coach-house, and call on me when you need help."

The boys did immediately proceed to make themselves at home in the club rooms, which they surveyed with a pride greater than they ever had experienced before. The rooms, small as they were, seemed spacious and important, and gave the boys a grown-up and authoritative feeling.

"Do you want your camera now?" asked Allan, handing the Wizard to McConnell.

"But I haven't the money yet," faltered McConnell.

"What of that? I don't need it. Whenever you are ready."

McConnell murmured a "Thank you, Allan," and gave the Wizard a friendly hug.

"And now," said Owen, "before we go any further, I nominate for President of the club, Mr. Allan Hartel."

"Hold on, Owen!" in[illegible], "you are the oldest, you must be [illegible]

Owen ignored this interruption. "All those in favor will please say, Aye!" and Owen and McConnell roared a tremendous "Aye!"

"It's carried unanimously!" exclaimed Owen.

"But, Owen — "

"And I move further," went on Owen, "that Mr. Percy McConnell be Secretary. All in favor will please say, Aye!" and Allan now joined with Owen in a resounding affirmative vote. McConnell looked overpowered.

"Then you must be Treasurer, Owen," declared Allan.

"Then elect me unanimously," said Owen, "or there will be trouble!"

"All in favor — " Allan began.

"Aye!" shouted McConnell, Allan following after.

"All unanimously elected!" said Owen. "That's good. Now we *are* a club!"

X.

THE CAMERA CLUB.

"I DON'T think," said Allan that afternoon, "that any club ever had more officers in proportion to its membership than this has."

"Yes," said Owen, "if we went into battle now the number of officers killed would be simply awful."

"I suppose Mebley will want to come in," said McConnell.

"And Varner," added Allan.

"And May Pelwin," Owen ventured.

"Will there be girls here?" asked McConnell, with perhaps a little of disappointment in his tone.

"Why not?" demanded Allan.

"Oh," said McConnell, with a shrug of his shoulders, "I didn't think they had girls in a club. I knew they had them in societies, and associations, and lodges, and circles, — but not in clubs."

"Pshaw!" said Owen, "girls belong to everything nowadays."

"I don't suppose we need worry," said Allan, "they may not think it is nice enough to join. We've got to fix this place up a good deal before many will want to join. Don't you think we ought to paint it a little?"

"I think we ought to clean it a little anyway," was Owen's opinion, "clear out the dust from the front room. You've got the dark room all right. Suppose we have another tap and sink, so that two of us can work at the same time if we want to?"

"Good idea," assented Allan, "and an extra lamp. Then we will want another fixing-tray — a large one; another rack; and I was looking at that washing-box Wincher has; what do you say to that?"

"I have been thinking," said Owen, "that we might have a gas meter put in and use these gas connections. Then we could run a line of pipe over the two sinks, and use gas in sliding-front red-glass boxes or something of that sort."

Allan and McConnell both thought this was a capital idea. "And what do you think," asked Allan, "of boxing in the head of the stairs, so as to keep out the light coming from the door in the daytime? We should have to do that anyhow to keep it warm here in winter."

"That's so," said Owen.

Allan drew from his pocket a piece of paper covered with pencil lines. "This is what I really was thinking of doing here," he said. The boys studied the diagram in which Allan had planned certain improvements in the dark room, and had set off the smaller of the two rooms on the front for a printing room.

"Just the thing!" was Owen's comment.

Yet, after much discussion, which occupied most of the following morning, the boys made radical changes in these plans. They finally decided to divide with a partition the room which had been used for developing, leaving access from the stairs to the front room without interference with the dark room work; and by making the dark room smaller several advantages seemed to be gained.

Unfortunately the estimates for these changes, from the plumber and carpenter, reached forty-five dollars.

"Then we can do the carpentering ourselves," said Allan. "That will make quite a difference." Upon inquiry they found this would make ten dollars difference.

"And we will have the initiation fees," said Owen.

They had decided on an initiation fee of three dollars for new members, with monthly dues of fifty cents.

"We have four applications," said McConnell. "That will make twelve dollars."

"Then we are safe enough," said Owen, "for there will be other new members as soon as we let it be known that we have a camera club with rooms here."

The week following was a busy one for the boys. It was early September, and the weather was warm; so warm that the amateur carpenters in the coach-house found their work very arduous at times, and were not unhappy when they were compelled to wait occasionally in order to keep out of the way of the plumbers.

Before the work was finished there were three more applications for membership.

"Say!" cried McConnell, one afternoon, "guess

who wants to be a member? Oh, you never will — Mr. Thornton!"

Mr. Thornton was the principal of the high school.

Allan dropped his hammer. "Mr. Thornton!"

"Yes. He says he thinks it is a good idea, and that he has been wanting a place where he could do his own developing if he wanted to."

"Mr. Thornton boards at Mrs. Peckpole's," said Owen, "and you know what a crank *she* is. Nobody could ever do any developing in *her* house."

"I never thought of men and women joining the club," said Allan, a little perplexed.

McConnell was decidedly amused. "What will they think of us kids for officers?"

"We shall have to have a regular constitution with by-laws," said Allan, "and then they can elect new officers."

What actually happened was this: Within two weeks there were twenty applications for membership in the club. The applicants included Miss Manston, the Mayor's daughter, Mrs. Creigh, the librarian, Mr. Austin, the Congregational clergyman, Mr. Goodstone of the bank, and Major Mines from the Ardmore Farm, — an aggregation that filled Allan with no little trepidation. When a constitution and by-laws had been drawn up, a thing happened that very much surprised Allan and his first associates; for on motion of Mr. Thornton all of the original officers were reëlected for a year. Allan could scarcely believe the vote, and Owen grew very red with embarrassment. As for McConnell, he seemed perfectly at home, and only chuckled with pleasure as he recorded the vote in his note-book.

"The first meet took place on the following Saturday."

The first formal meeting had been held in the larger of the two front rooms in the coach-house, and the fifteen members who attended did not find themselves greatly crowded. Under the new constitution two officers were added — a Vice-President and a Curator. Mrs. Creigh was elected Vice-President, and the office of Curator was bestowed upon Mr. Goodstone, who, it appeared after a while (the boys could not fancy, at first, what a curator was for), would supervise the buying of supplies for the club, and have authority over the club rooms. The funds from the increased membership enabled the club to complete in a satisfactory way the arrangement of the dark room and other quarters. One feature of the small room created by the dark room partition was a series of lockers, each member thus being provided with a place in which to keep his personal implements and supplies.

Dr. Hartel refused to accept any rental for the rooms, but did not refuse an election as honorary member, and watched the progress of the club with great pride. When Edith read in the Hazenfield *Herald* about the Camera Club election of officers, and saw Allan's name at the head of the list, she laughed with pleasure.

It was at the first meeting that Mr. Austin had said, "We must have an outing. There will be but a few more weeks of good outing weather."

The proposal met with favor. Mr. Goodstone suggested Saturday afternoons for weekly club "meets," and the first meet took place on the following Saturday, when twelve of the members, young and old, went on a short tramp into the hills.

Allan thought that nothing about the trip was so interesting, so truly picturesque, as the club itself; for it was a mixed company truly. The most serious-minded photographer in the group was Major Mines, whose hired man, Napoleon, a portly negro, carried an immense outfit, bristling with shining modern improvements, heavy with conveniences, and packed into the corners with things you might want. Some of the boys were always ready to help the Major get Napoleon and his outfit over a fence; and, indeed, such assistance was necessary if the Major was to make any progress. The Major made two pictures during the day, and for these he made elaborate preparations, choosing his view-point after long study on a plan of action he had laid out, and setting up his camera only after long wrestlings with the many improvements and conveniences that were stowed in his carrying case. His bald head remained under the focussing cloth for minutes at a time. Twice he hobbled down a field to break off discordant sprigs and branches. When he made the exposures his face was as tense and solemn as if he was giving the signal at an execution.

A little Miss Illwin also took her photography with great solemnity. Miss Illwin was very cautious, too. She did not believe in wasting plates. All the afternoon she debated about a good point of view. She scrutinized the spots selected by other members, and then shook her head. "The light doesn't seem quite right," she said. Miss Illwin had a small, dainty camera, and she studied the finder frequently, puckering her white forehead and shifting her eye-glasses in an earnest and tireless way. At last toward four o'clock she was discovered on the brow of a low hill overlooking a brook. For a long time she stood there in the sun, quite motionless, with her head under the

"Into the hills."

focussing cloth. Then her head lifted, a plate holder was placed in position, the exposure accomplished, and she turned and hurried after the nearest group of members.

"I made an exposure over there," she said, with something like a sigh, "but I don't see *why* I did it. It would have been much better at the foot of the hill!"

Mr. Goodstone and Varner and Mebley were all very exact and painstaking, but they took as much pleasure in shooting at one thing as another. Miss Manston frequently asked Allan's assistance here and at her house. Mrs. Creigh was enthusiastic over many things, and worked with much ardor under a heavy, satin-lined focussing cloth. Young Coggshall complained during the whole trip, — of the heat of the sun, of the stupidity of the landscapes, of the meanness of plate-holders, of the superiority of other landscapes, other times of the year, other kinds of plate-holders.

Mr. Austin was a very different companion. He watched Allan and the boys at work, encouraged Mrs. Creigh, gave a hand to Major Mines, assisted Mr. Goodstone in a tape-measuring experiment, led an expedition to a hillside farmhouse after a pail of milk, and joined McConnell in a camera duel, Owen standing by and dropping a handkerchief as a signal to fire.

In the suburbs of Hazenfield Mr. Austin found a group of barefooted youngsters, a subject entirely to his taste — no one excelled Mr. Austin in pictures of children — and while he was photographing the group strung along in the road, Allan caught both group and photographer, laughingly turning his head when he had done so to see whether any one had caught him in turn. This happened many times during the day, that the picture-makers were pictured. Indeed, Allan's pictures were almost entirely of the club itself.

The same thing was true of the club's trip along the river front, where shore and river offered so many interesting backgrounds to any theme the camera might select. On this day McConnell made a great success with a train coming through a cutting.

It was at Allan's suggestion that the club decided to spend a Saturday at Central Park in New York, and this proved to be one of the most entertaining and successful trips of the club. In the first place nineteen members mustered in the morning, which was regarded as a large attendance; and Mr. Goodstone, who was kept at the bank in the morning, hunted up the party before the afternoon was far advanced.

"This puts me in mind of a Cook's tourist party," said Mr. Austin.

"With the Major 'personally conducting' us," laughed Mrs. Creigh.

The Critical Moment.

"A group of barefooted youngsters."

The fact is that while Owen, as treasurer, had attended to getting the tickets at the station, and Allan, after consultation with Mr. Thornton, had written to the office of the Park Board for permits for the club, the Major, by general consent, had been selected to lead the party in the park, he being most familiar with all the features of this beauty spot in the heart of Manhattan.

The lakes and bridges and grottoes of the park never were more ardently photographed than on the day when the Hazenfield Camera Club descended upon it. Even the big white bear at the Zoo, who always looked bored when he saw a camera, stared curiously at the Hazenfield party ; the ostriches looked scornful, and the rhinoceros seemed likely to strangle himself in an effort to keep himself out of sight under water.

Of course Allan photographed the elephants. All the members wanted to photograph the elephants, and Major Mines induced the keeper to make special dis-

"The big white bear."

play of the Princess; and the Princess, while not at all guilty of looking pleasant, at least turned an almost motionless profile to the bristling battery of cameras.

"Mercy!" cried Miss Manston, "I never should have supposed anything was so hard as photographing an elephant with so many people looking on!"

There were a great many onlookers, and many of

them felt quite free to comment, not merely on the elephant and the camera, but on the photographers. Miss Illwin, with her head under the focussing cloth, was an object of much interest. Miss Illwin had a little loop stitched to her hat so that she could suspend it from a hook on the under side of the tripod after it had been set up. This left her free to study

"All the members wanted to photograph the elephants."

her ground glass without greater disturbance than the mussing of her hair, which did not seem to annoy her at all.

It was a balmy day, and the out-door cages were full of listless and sleepy lions, tigers, leopards, wild-cats; bison, zebras, camels, and deer roamed in the inclosures; the eagles screamed, and the monkeys were in their most talkative mood. Photographing through the bars was a delicate problem. The pelican

"Photographing through the bars was a delicate problem."

and other queer birds strutted and squeaked and flapped their wings at the visitors and at each other. The children who peered between the bars of the cages, who laughed with the parrots, threw peanuts to the monkeys, or stared in awe at the dromedary, were a camera theme in themselves; and Mr. Austin often was seen to be picturing or talking to them, or slipping pennies into their hands on the outskirts of the crowd at the candy stall.

Allan and McConnell were commenting on the mountain goat, and McConnell was saying, "He has a Van Dyck beard, hasn't he?" when Allan caught sight of a man over by the zebras, who was studying the finder of a hand camera, and thoughtfully puckering his mouth until his mustache looked more than ever like the bristles of a brush.

Allan left McConnell and ran toward the man.

"Queer birds."

"Hello, Mr. Dobbs!"

"Well, I'll be blowed!" exclaimed the man; "where did you come from?"

"You're a nice detective," laughed Allan; "here's half of Hazenfield, and you haven't seen us!"

Dobbs grinned. "I just got here." Then he held up his camera. "I had to get one, and I sneaked up here to try it. The worst crook in the country would be perfectly safe to-day — I haven't been able to see anything but this finder since ten o'clock. But what is the crowd doing here? why, there's Major Mines, and Mr. Thornton — and Goodstone."

"This is the club," said Allan.

"The club?"

"Yes; haven't you heard of it? — the Hazenfield Camera Club."

"Why, yes," Dobbs said. "I did hear something

about a club, but I thought it was only two or three of you boys."

"It has grown since then."

"Evidently—hello, McConnell! You're in it too. Say, I want to be a member, if you'll take me."

"I don't see why not," laughed Allan, "now that you're a photographer."

"Well," said Dobbs, moving the bristles of his mustache again, "I'm not much of a photographer yet. I've only had courage to push the button once. I was just going to take the zebra—thought Sporty might like to see it."

"What kind of a camera is yours?" asked McConnell.

"Just a plain Detective, I suppose," interposed Allan, laughing.

"It's a Dashaway," said Dobbs. "I got it through a pard of mine here in New York. He says it's a good one. Sporty and me'll have great fun with it. What do I have to do to get into your club?"

"That's easy," McConnell said. "Three dollars initiation fee and fifty cents a month."

"It's too cheap," said Dobbs. "How do you get initiated?—in the dark, of course."

"Yes," said Allan; "you have to sit on a tripod, focus a camera over your left shoulder, and recite the eiko-hydro developer backwards at the same time."

"Or treat?" asked Dobbs.

"Or treat to a bottle of developer."

"I'll risk it," said Dobbs, "if you'll let me in."

Mr. Thornton came up and recognized the detective. "Good afternoon, Mr. Dobbs; looking for the owner of a camera?"

"No," returned Dobbs; "I've found him. This is mine, and I want to join your club if you'll let me."

"The more the merrier, Mr. Dobbs. You must join us to-day, anyhow. They are not fitting out the detective force with cameras, are they?"

"Not yet. There is no business about this. I am playing truant to-day."

"I dare say you deserve a holiday," said Mr. Thornton.

"Well, I've been tied down awfully close to that Bain case. Glad it's over." Dobbs turned to Allan. "You remember the Ghost?"

"Yes."

"Well, we sent him up yesterday for fifteen years."

"Fifteen years!" Allan recalled the white face of the man at the police station. It was the first time that any one he had ever seen had been sent to prison. Fifteen years! It seemed like a sentence for life. Could that white face grow any whiter in fifteen years?

"He was a hard case," said Mr. Thornton.

"He was that," said Dobbs. "As bad as they make them. But a queer fellow; you never could make him out."

"What did he say when they sentenced him?" Allan asked of Dobbs.

"Not a word, you couldn't get anything out of him. Even ghosts speak, they say. But this one was silent as the grave."

"Did you take the zebra?" broke in McConnell.

"No," said Dobbs. "I was just flirting with him. But I think I will take him."

It ended by their all following the zebra, who looked very much amused, and finally came over to push his nose through the wires.

"He isn't painted, either," said McConnell, scratching the zebra's back.

"Oh, this is a very honest show!" laughed the detective.

It turned out that Dobbs knew a good many of the people at the Zoo, and before the company started homeward he had made it possible for the club to do pretty much as it pleased.

When the club was ready to start, McConnell found every one but Miss Illwin. Owen had seen her over by the rhinoceros tank. Mr. Goodstone and Mrs. Creigh had left her with the deer. But no one was able to actually find her.

"Where can she be?" queried Miss Manston.

"You don't suppose anything could have happened, do you?" asked Mrs. Creigh, her face indicating real anxiety.

"Well," said Major Mines, mischievously, "she was taking the tiger at twenty feet. It seemed safe enough."

"The lady or the tiger," muttered Mr. Thornton.

"I really think," said Mr. Goodstone, solemnly, "that some of us ought to look up the rhinoceros part of the story."

"Do you think so?" asked Miss Manston, half inclined to think this was no joke. She was so afraid of rhinoceroses herself. "Horrors! Suppose she fell in!"

"Don't!" protested Mrs. Creigh with a shiver.

McConnell and Allan went out as a scouting party, with the result that Miss Illwin was found sitting on a bench by the lake, her camera carrying case beside her. She was reading a book. As he came up Allan noticed that she was without her hat.

"You know," she said, when she saw Allan, "my hat blew into the water when I was setting up just now. Wasn't it annoying? I was afraid I should have to wait until it drifted over, but that boy with the little yacht aimed his boat so nicely that it caught the hat not far from the shore and now it is pushing it over. See!" and Miss Illwin pointed toward the middle of the pond. "Unfortunately the wind has shifted once or twice, and the yacht has been tacking about in a most provoking way. But there is nothing but to be patient."

"We are all ready to go," said Allan.

"Are you? Then I suppose you'll have to leave me — unless this boat stops tacking."

"There it comes!" yelled the boys who owned the boat.

The yacht was making straight for the shore. But when it had come within about fifteen feet of the bank, the band of the hat became loosened from the bow, and the yacht came jauntily into port, leaving the hat in its wake.

"Dear me!" exclaimed Miss Illwin. "How provoking."

"The breeze is carrying it, anyway," said McConnell.

But the hat drifted very slowly, and finally stopped altogether, anchored by a water-lily.

"If you will let me have your tripod, Miss Illwin," said Allan, "I think I can reach it with that.

A single leg of the tripod proved insufficient, but, by hitching two of them to the top piece, Allan managed to reach the vagrant hat.

"Is it spoiled?" asked McConnell.

Miss Illwin shook the bedraggled hat. "I'm afraid

it is," she said, adjusting it with a wince, and pushing through the pins. "What a providence that I didn't wear the one with the feather!"

The club appeared much relieved to see Miss Illwin.

Everybody had a story about the loss of a hat, but no one claimed to have recovered one under such picturesque conditions.

"Well," said the Major, as they walked toward the "L" station, "I'll wager none of you ever had your hat blow into the crater of a volcano."

"Gracious! Suppose you had gone with it!" cried Miss Manston.

"Couldn't you get it with a pole, or a hook and line?" asked McConnell.

"No, I couldn't," grunted the Major. "Never saw the crater of Vesuvius, did you?"

"No," admitted McConnell.

"Well, it isn't good fishing there."

"Fancy!" exclaimed Mrs. Creigh.

"And I had to go back bareheaded — I refused a crazy bonnet that one of the guides found for me — and those people at the hotel said, 'These Americans are so funny!'"

"Where shall the club go next!" some one asked.

"Better wait until you find out what sins you committed to-day," said Mr. Thornton.

XI.

AT CONEY ISLAND.

THE scene at the club that night was a busy one. Mrs. Creigh and Mr. Thornton were first to begin developing, and several other members who did not have dark rooms at home came around and waited their turn.

Dr. Hartel had advised Allan not to do any developing under any circumstances, but to wait until the following day, when everything was quiet. Allan thought that anyway it was but fair for him, living so near, to make use of the dark room at times when there was not a pressure of other members.

It was the next morning that Allan and McConnell developed their plates and films. Allan had two wrongly focussed pictures, and McConnell had made a double that perched a goat on the back of a dromedary. But most of the pictures were good, and gave both boys a great deal of delight.

"Wait till Bill sees this!" exclaimed McConnell, holding up his picture of the lion cage.

Allan highly prized a picture of the eagles, and another of the club emerging from one of the arches under the drive.

"Do you know," said Allan, "the club won't be taking another outing for a couple of weeks. I wish we could go to Coney Island before that. If we wait two weeks they might not want to go there, and I'm afraid if the weather gets cooler the season will be over in a week."

"What do we care for the season?" asked McConnell.

"The season doesn't make any difference to the ocean, — except that it's the clearer, I suppose, when so many people don't wash in it; but it makes a great deal of difference at the beach. It's the people I care for, the people and all the things they have there to show the people."

"That's so," said McConnell.

"I wonder if Owen would go?"

"Of course he would. I guess my mother would let me go with you."

"Suppose we go on Wednesday. School begins next week."

"How much would it cost?" asked McConnell.

Allan figured the thing out on the back of a plate-box. "If we carry part of our lunch with us it ought to cost about seventy cents for each of us."

"And we could get some great things down there," commented McConnell. "We could shoot from the Ferris wheel."

"Yes, and we could shoot the chutes."

McConnell chuckled. "I'd like to go."

Owen liked the idea. "I've been wanting to go down all summer. And what do you say to this? There is a freight-boat that stops at Howlett's Dock every morning about seven and comes back every night. I know one of the men on it, and I think we could get him to take us down to Twenty-something Street and back. That would only leave us the Coney Island boat to pay for."

"Good!" cried Allan.

Owen's plan was carried out. Wednesday morning opened cloudy, and the boys were not in very high spirits when they reached the rendezvous at Howlett's Dock.

"Is there any use going when it's so cloudy?" asked Owen.

McConnell was certain it would clear.

A man was fishing from the end of the dock. Allan went over and asked him if he thought it was going to rain.

"Rain before seven, clear before eleven," said the man, without looking up.

"But it isn't raining," chimed in McConnell.

"Yes, it is," said the man, without moving; and then Allan felt a drop on his hand.

"What a strange man," said McConnell, as they moved away.

"When they are queer like that they always know about the weather," said Allan.

"What did he say?" asked Owen.

"That it would clear before eleven."

"Then that's all we want. We won't get to Coney Island much before twelve."

The fisherman was right. The light drizzle fell until after nine. At ten the clouds began to lift; and while they were on their way down the bay on the Coney Island boat, the sun came forth cheerily.

Allan celebrated the arrival of the sunlight by photographing the group of Italian musicians on the

"The group of Italian musicians."

upper deck, and McConnell blazed away with the Wizard at an ocean steamer that had just come over the bar.

An old gentleman in a bicycle suit who sat smoking at the forward rail, after watching the three boys with cameras, fell to talking with them, and soon had examined all three apparatuses in a way that indicated some knowledge of photography.

"These are all very convenient," he said. "They make me want to take up photography again."

"Did you once have a camera?" asked Allan.

"Yes, when I was younger I had a wet-plate outfit. I don't suppose you know what that was?"

The boys shook their heads.

"Well, in the wet-plate days, before there were any dry plates such as you use, and very long before there were any films," the old gentleman added, indicating Allan's Kodak, "we had to coat and sensitize our own plates before exposing them. We had to do this where we made the picture, or very near it, for we had to expose the plate while the coating was still somewhat moist."

"It must have been a lot of trouble," said McConnell.

"It was; but I enjoyed it."

"Then you had to develop them right away, too?" suggested Allan.

"Yes, we had to do the whole thing — sensitize, expose, and develop — at the one time. Ah!" continued the old gentleman, slapping his knee, "but they made beautiful plates! More beautiful than your new-fangled plates!"

They were now nearing the iron pier at Coney Island.

"Boys," said the old gentleman, "what are you going to do to-day?"

"I don't know that we are quite sure," laughed Allan. "We have carried our cameras down to shoot at the island — anything that seems interesting."

"I tell you what I wish," continued the old gentleman. "I wish you would go up in the Ferris wheel with me."

"Thank you," replied Allan, "we did want to go up in the wheel."

Owen and McConnell indicated their willingness.

"I've been around the world since I saw Coney Island before," the old gentleman went on, "and I have run down to see how much it has changed. I suppose it has changed a good deal. I have too." The old gentleman smoked in silence until it was time for them to go ashore.

The beach was not so crowded as on a Saturday, but there were animated scenes on every hand. A great chorus of sounds went up from the West End — the shouts of hawkers and doorkeepers; the blare of a dozen merry-go-round organs; the whir and clatter of the switch-back railways; the hum of thousands of voices; the screams of children at the water's edge, mingling with the swish and roar of the surf.

"Just the same!" said the old gentleman, smiling; "only more so!"

The old gentleman led the way to the largest of the wheels that swung its great spokes into the air. Allan took a seat with the old gentleman in one car, while

"The screams of children at the water's edge."

"The largest of the wheels."

Owen and McConnell stepped into another, when the engineer had swung it into position.

"I suppose they distribute us this way to balance it," said Owen. The wheel now began to revolve slowly.

"My name is Prenwood," said the old gentleman. "What is yours?"

"Hartel," said Allan. He already had said that the boys were from Hazenfield.

As their car swung over the top of the wheel Mr. Prenwood exclaimed, "Yes, the same old place!" and was silent again for the whole circuit of the wheel. Allan found it harder than he expected to accomplish a sighting of his camera. There was something curiously confusing in the constantly changing situation of the car. Looking for the other boys was also a difficult matter. Sometimes they were below him, sometimes above. McConnell was making persistent efforts to bring his Wizard to bear on the beach without having the rim or spokes of the wheel in the way. Allan could hear him laughing, and Owen urging him not to lean over so far unless he wished to fall out.

"Hartel," said Mr. Prenwood, "I used to come down here with my little nephew. He was too young to see any of the vulgarity. He just enjoyed the life and stir, the bustle of the place, just as you boys do. He called these wheels — they were little then — the big pin wheels; and those were the 'slam-bang railroads.' It was fun to watch him! While I was away in the Mediterranean they buried him." Mr. Prenwood sat very quietly for a moment. "They tell me that last summer he used to say he wished Uncle Amos would come home, and take him to Coney Island. No one else would take him, it seems. And now no one can take him. Isn't it a shame, Hartel, that I couldn't have been here?"

"How old was he?" asked Allan, who could see tears in Mr. Prenwood's eyes.

"Only six," said the old gentleman. "What a little man he was! You should have seen us wading on the beach, and how he used to laugh when I rolled up my trousers! And he seemed to know just how funny it was when I sat on a horse beside him in the merry-go-round. But we are missing the view altogether. How gay the sun makes everything look! What a good thing it is the sun never gets sad! If the clouds will only let him shine, he's always as jolly as ever."

The sun shone on the big, creaking wheel. Mr. Prenwood waved his hand to Owen and McConnell. A young girl who was sitting beside a young man in the car at the other side of the wheel seemed to think the salute was intended for her, and giggled.

"Why not take the wheel, with the couple at the other end?" laughed Mr. Prenwood.

Allan already was preparing to do this. The spokes and cross-bars made a curious cobweb of lines in the finder, a cobweb that twisted like a kaleidoscope.

When they all had stepped out of the wheel again Mr. Prenwood said: "Now, boys, I'm not going to bother you with my company much longer. You have things that you want to do on your own account. But I would like very much if you would go with me and have lunch. When I was a boy I got frightfully hungry at this time of day, and I haven't altogether recovered from the habit yet."

Owen thanked Mr. Prenwood, but said they had some lunch with them.

"Yes, I know," laughed Mr. Prenwood, "you have some dry lunch. But that isn't enough. Oh, I know! Come along. We'll eat that — and something else with it! Some oysters, for instance, or chowder, or sweet corn, or watermelon. And how about buttermilk and pie and ice-cream? Hey?"

And they all laughed together as they followed Mr. Prenwood up the broad walk.

"I tell you," said Mr. Prenwood, laughing again, as they sat at a table in the hotel dining-room, "those dry lunches are well enough to start in with. But to finish up on! — Waiter! ask these boys what they wish, — and boys! see that you wish everything nice on the bill!"

While they were eating their unpacked lunch and many good things brought by the waiter, Mr. Prenwood told them some of his adventures in the days of wet-plate photography. "You boys must come over and visit me sometime. I haven't told you yet that I live on the other side of the river — at Stonyshore. I've got dogs, horses, boats, — everything but boys. I've half a mind to steal one of you!"

When they were shaking hands and saying good-by to Mr. Prenwood, he told them not to forget to come and see him, and he saw in their faces that they were very glad they had met him.

"Wasn't he nice!" exclaimed McConnell, when they had walked away.

"I think we ought to send him over some prints," said Allan.

"I believe I will," said Owen, "if I get anything good."

Later in the day they saw Mr. Prenwood sitting on a bench, smoking, near one of the merry-go-rounds.

Immediately after their luncheon the boys went down on the beach and walked the whole length from Brighton to the far West End. The strollers, the children wading in the foam, the sleeping figures in the sand; the chair men wrangling over the price of seats; the chowder boats and ring-toss tents; the bathing-houses, and screaming bathers in the surf; the groups at the photograph galleries, — these and a score of other sights gave the boys amusement and an embarrassment of themes for their cameras.

Then they went over to the chutes and found it

"Wading in the foam."

more exciting to try and photograph the flying boat than they had found it to ride in one. Allan wished to catch the boat just as it left the incline and struck the water. When he had developed his plate he found that the spray hid the boat.

They walked through Coney Island's Bowery, but it did not please them. They liked the beach better. There was much fun walking in the sand. The three boys for a while played quoits with clam-shells. Far up at the West End two young men were having a wrestling match and insisted upon being photographed while they were at it.

"Our style of wrestling is not set down in the books," said one of the young men. But they had great fun at it; and all hands had a hundred yard dash afterward. The taller wrestler came in first and Allan second.

The winner afterward said to Allan, "I'd like to

"A wrestling match."

have a copy of that wrestling picture, and if you'll promise to send me one I'll give you this pass to Buffalo Bill."

"But perhaps it won't be good — I mean the picture."

"Then you're so much ahead. I'll take my chances." And he gave Allan his name and address.

The pass was for two. "Isn't he coming to Granger Fields next week?" asked Allan.

"Yes. If you live near there, that's the place to see him. And take your camera, too. Say; wait a minute. Would you like to do the Indians? Well, you ask for Mr. Twink — he'll fix it so's you can photograph the Indians. Tell him I sent you. He's my cousin. I know the whole crowd pretty well."

"The groups at the photograph galleries."

"I'll do the best I can with the picture," said Allan.

"And I'll take what I get and be thankful," laughed the young man.

This seemed decidedly like a stroke of luck to Allan, not so much for the value of the pass as for the chance to get special privileges with his camera — with their cameras, for undoubtedly Mr. Twink would look with a friendly eye on any party that might come with his cousin's name.

"Time for the boat!" cried Owen.

The day had slid away so quickly and they had given so little thought to time that there was not a moment to be lost if they were to catch the 4.30 boat.

The two started off for the pier on a run, until the little brown woman in the Turkish bazaar looked up from her beads to see whether any new and special excitement had befallen her street. A short cut across the sand proved to be heavy travelling, and the boat's whistle sounded warningly in their ears. All three were much in need of breath when they reached the pier. They caught the boat. But it was a narrow escape.

XII.

BIG WOLF AND COMPANY.

BUFFALO BILL was to be at Granger Fields for three days. Great crowds came to see the Indians, and the Oriental acrobats, and the soldiers of many nations, and the "rough riders of the world." On the third day, half an hour before the time for the beginning of the performance, Allan and McConnell arrived at the gate.

Up to the last moment it was expected that Owen would join them. But Owen had been unable to come for some reason, and the two boys had walked the three miles alone. The pass carried them past the ticket man and admitted them to the grand stand. But before taking seats there, the boys started out to find Mr. Twink. A sleek little young man, with long hair and a big sombrero, who looked like a candy cowboy, told them where they would probably find Mr. Twink.

Mr. Twink, however, was very hard to find. Each person they asked said he was in a different place.

When they came down by the Indian tents, they at last found Mr. Twink. He was talking to an Indian, — an Indian decked out in gaudy red and yellow, and with many feathers dangling down his back.

When they at last got his attention, Mr. Twink told the boys that everything was in a hurry just then, that the Indians were getting ready for the grand entrée; but that if they would come around after the show, he would give them a chance to photograph all they wanted to.

"Hold on a minute!" he called after them, as they turned away. "We are going to strike camp this afternoon. You had better come around as soon as the acrobats begin. You can get back in time to see the cowboy and Indian fight."

"Where shall we find you?" asked Allan.

"Right here," said Mr. Twink. "Don't forget — as soon as the acrobats begin."

Allan promised to be prompt.

"But I hate to miss the acrobats," said McConnell, regretfully.

"We can see acrobats any time," protested Allan, "and we may never get a chance to photograph Indians — close up — again."

They both photographed the grand entrée from their seats in the grand stand, though the figures of the soldiers and cowboys and Indians and Arabs looked very small at that distance, and the heads of the people in the spectators' seats made a rather conspicuous foreground. They caught the bucking broncho while two of the cowboys were trying to master him.

When the Arab acrobats came out, the boys slipped out of their seats and went around to look for Mr. Twink. He was where he had said he would be.

"Now, what do you want to do?" he demanded, so abruptly that Allan was a little at a loss what to say.

"We should like to photograph some of the Indians," said Allan, finally.

"Well, here's Walking Dog, photograph him." And Twink caught a passing Indian by the arm.

Walking Dog was very solemn in appearance, and when Twink said something to him in a language the boys could not understand — it was the first Indian talk Allan or McConnell ever had heard — Walking Dog looked at the boys and at their camera without a smile.

Allan was sure that Walking Dog resented the proposition to be photographed, and felt sorry he had mentioned it. The Indian looked so savage in his paint.

"He says all right," remarked Twink.

Now, Allan was sure Walking Dog had not uttered a sound, and he wondered very much what language the Indian had used that Twink should feel so sure.

"How about taking him over here?" said Twink, pointing to a spot where a stretch of canvas would form a background.

Walking Dog seemed to understand at once, and, striding across, he stood with his back against the canvas, his hands on his rifle, and in a position such as soldiers take at "parade rest."

Walking Dog refused to look pleasant while Allan and McConnell got their camera ready. Or perhaps it was his natural look with the ugly war-paint added.

"Get it?" asked Twink, when he heard the camera click.

"Oh, wait a minute!" cried McConnell. He had forgotten the slide of his plate-holder. Allan rolled his cartridge another number, and took one more to keep McConnell company. Walking Dog remained as still as a soldier's monument while this was going on.

"Walking Dog refused to look pleasant."

"I wish you would tell him we are much obliged," said Allan to Twink.

"Oh, he knows that," said Twink; but he spoke to the Indian as he was moving away, and Walking Dog shook hands with both boys, stiffly and silently, then walked majestically away.

Twink now left them for a moment and spoke to another Indian, a much handsomer and more gorgeous Indian, though one not less solemn than Walking Dog. They came back together.

"This is Big Wolf," said Twink. "You can take him, too."

"Ugh!" said Big Wolf, and he sat down near by, looking straight before him.

"Don't they like to do this?" asked McConnell. He couldn't get it out of his head that Big Wolf was likely to rebel at any moment and scalp them both.

"Oh, they are not so bad as they look," said Twink, smiling for the first time. "They are quite sociable when you come to know them."

"Is Big Wolf a chief?" asked Allan.

"Or just — just a plain Indian?" added McConnell. It seemed incredible to both boys that Big Wolf could be less than a very important personage.

Twink waved his hand. "Big Wolf heap big chief!"

"Ugh!" said Big Wolf.

"See — he admits it himself," said Twink.

The boys did not dare to smile. It would have seemed very inappropriate with Big Wolf sitting there so solemnly.

When the pictures had been taken the Indian arose and left them, giving a quick nod to Allan as he went by.

"Now," said Twink, "if you hurry I think I can get up a group for you. Come over here."

The boys followed their guide across the field to where several of the Indian tents were grouped. On their way over Twink said, "There's Buffalo Bill."

Colonel Cody was seated near a screen of canvas at a point where he could watch the arena through a hole that had been cut for the purpose. The boys had no time to look closely at the famous plainsman, for Twink was hurrying them over to the tents. Twink spoke to several of the Indians, and presently, before one of the tents, a line of Indians was formed, a squaw and baby at one end of the line. These Indians had shields and other weapons, and stood bolt upright in all their gay colors, and waited without sign or sound while Allan and McConnell each made two shots with their cameras.

"If you come back here after the show," said Twink, "you can see them striking these tents. Meanwhile, make yourselves at home. Here is the famous old

"The Deadwood Stage-coach."

Deadwood Stage-coach. Would you like to ride in it to-day?"

"Oh, yes!" answered McConnell.

"Well, when the coach draws up at the grand stand you boys just climb down and get in. I will speak to them about it."

"Will anybody take us down?" asked Allan, a little uncertain about the programme.

"A line of Indians."

"No. You just get down yourself. The men in the coach will be watching for you."

They stood looking at the battered old stage-coach after Twink had left them, and a man with a coat on his arm told them that one day, in England, when the show was over there, a king and four princes had ridden on it, Buffalo Bill himself driving.

This nearly took McConnell's breath away.

"Do you know where the king sat?" asked McConnell.

"I dunno," said the man. "I guess up beside Buffalo Bill."

"Where the princes sat ought to be good enough for us," laughed Allan. "Come," he added, "let us go back to the show," and they hurried around to the grand stand in time to see the Mexican throwing the lasso.

"I don't believe we thanked Mr. Twink," said Allan.

"Won't we see him again?" asked McConnell.

"That's so. We can see him before we leave. I feel as if he had been very good to us."

"Indeed he has."

"We never could have seen so much and got those Indians without him."

"Do you suppose he would care for a picture of Big Wolf?"

"I don't think he would."

"We might ask him."

"I almost think we had better not bother him again."

"Except to thank him, you mean?"

"Except to thank him, yes."

Presently, the man with the wonderful voice, who

"The coach itself, drawn by four horses."

made the announcements, told how the great Deadwood stage-coach would be attacked by the Indians, and how it would be rescued by a company of cowboys under the leadership of Buffalo Bill.

The coach itself, drawn by four horses, now came rolling around the arena.

"Are you going?" asked McConnell, his eyes twitching with excitement.

"Of course," replied Allan; but he could not have concealed his nervousness.

As the coach drew nearer the grand stand the boys rose and clambered down the steps to the main entrance; and, when the coach stopped, they walked falteringly forward, expecting the man at the bars to ask them what they wanted anyway.

But the man at the bars, on a signal from the coach, made way for them, and the old coach door opened. They now saw that there were two men on the front inside seat, and, with several thousand people watching them, the boys climbed in and sat down on the back seat, the door closed, and the coach started forward with a jolt.

The whip cracked and soon came the louder crack of a rifle, then a clatter of shots, and the two men in the coach, each with a rifle, began blazing away through the window at the yelling band of Indians in pursuit. It was all so real, the Indians looked so ferocious, the smoke and flame from the rifles was so thrilling and threatening, that Allan and McConnell found themselves shrinking in expectation of actual bullets.

In the midst of the hubbub Allan saw through the window, almost at his elbow, the now distorted face of B[illegible] Wolf, screaming a most frightful note, and a[illegible]ntly on the point at last of getting even with his pho[illegible]graphic tormentors.

Then came a new and louder clatter, with fresh yells. The cowboys had come, and, after a wild fusillade, the Indians fled, the smoke cleared away, and the old coach lumbered back to the grand stand, with Allan and McConnell staring, half-dazed, at the two men on the front seat.

"How did you like it?" asked one of the men, as he swung open the door.

"It was great!" cried Allan.

McConnell could hardly find his voice. "I guess it was like being in a battle!" he said, as he climbed out.

"Just like it," laughed one of the men, "only that you haven't got any lead in you!"

"Did you see Big Wolf?" asked Allan, as they walked, rather weak in the knees, back to their seats.

"Yes," answered McConnell, "and I thought I saw Walking Dog, but I wasn't sure. I suppose he was there. Wasn't the noise awful! — and the smoke! I can see now why they can't photograph well in a battle — unless they use that new smokeless powder I was reading about."

There was more of the show, but nothing seemed so thrilling as that ride. In the midst of the last performance McConnell leaned forward excitedly to say, "Suppose some of them forgot and put in real cartridges!"

When the Congress of Rough Riders had drawn up in line and Buffalo Bill had swung his big hat in a final salute, the boys once more hurried around to the Indian tents, and found the Indians all very busy in preparation for departure, and the wigwams gradually disappearing.

"The wigwams gradually disappearing."

Mr. Twink was nowhere to be seen, and nobody seemed to know where he was. It seemed for a time as if they would have to give him up.

"Oh, there's Big Wolf!" exclaimed McConnell. "I suppose he would know where he was."

"I've a mind to ask him," said Allan. In a moment he did gather courage to hurry over to where Big Wolf was standing, solemnly and deliberately folding a red blanket.

"Do you know where Mr. Twink is?" asked Allan, in a loud voice, as people always do when they talk to a foreigner or one whom they fear will not understand them.

Big Wolf turned and mutely pointed toward a distant group of men. Yes, Mr. Twink was there.

"Thank you," said Allan. Big Wolf went on folding the blanket.

When they got over to where Mr. Twink was, Allan caught his attention, and both boys stammered their thanks to him.

"Oh, that's all right," said Twink. "Glad to have given you a hand. Come and see us again sometime. Good-by!"

And they left him.

"It seems to me," said McConnell, as they walked home in the early evening, "it seems to me that wonderfully interesting things happen to you when you have a camera!"

"I was just thinking the same thing myself," said Allan, swinging his black box. "I don't suppose we ever should have thought of going 'behind the scenes' as we did to-day if we hadn't these cameras with us."

"And we couldn't have talked to the Indians," McConnell added in a tone of profound satisfaction.

"Well, I don't suppose we should have had any excuse."

"Yes," said McConnell, "the camera is an excuse, isn't it?"

XIII.

A TOUCH-DOWN.

McCONNELL'S remark now began to seem entirely true; for even commonplace scenes and commonplace happenings became more interesting than they ever before had seemed, now that they were associated with picture-making.

Dr. Hartel said that this was because the boys began to think about things in a new light, of which they never before had thought about at all. "It would be much the same," he said, "if you had taken up botany, or mineralogy, or the microscope. I remember that life and history and governments suddenly began to have an entirely new interest for me when I began collecting coins and postage-stamps. Before that it didn't seem to make much difference about the Italian States or the precise date of the Restoration, or who was restored. Then at once it began to seem of positively exciting importance. My stamps and coins began telling me when and whom. It is the same with your new hobby. When a man climbs on a hobby, unless he rides it too hard and loses his balance, he gets a wider view of something."

"So I mustn't ride too hard," said Allan.

"No, you must remember that the academy opens next week."

A week after the high school opened, McConnell told Allan that Mr. Thornton had remarked one afternoon, "Now, McConnell, I'm afraid you are thinking about your camera."

"I guess the next time I get him at one of our meetings," laughed McConnell, "I'll say to him, 'Now, Mr. Thornton, I'm afraid you are thinking about the high school.'"

At the Camera Club they had begun to talk about an exhibition, and a committee was appointed to talk the matter over. It would be a good idea, several of the members thought, to have a display of the summer work. Many of the members had travelled to the mountains and seashore during July and August, and an exhibition could be made to have great variety in theme. Moreover, the club excursions had produced a large batch of pictures, and the members had not yet seen much of one another's work.

"We might want to do some swapping," said Owen.

"I wonder if we could get Dobbs to exhibit," said Allan, amused at the thought. "I'm sure he would have something different from anybody else."

"I fancy Dobbs is having a hard time," said Owen. "He told me yesterday that he had taken six pictures of — what's his boy's name?"

"Sporty."

"Yes, Sporty, — I hope that isn't his real name, — and that not one of them came out good. He seemed disappointed."

Allan had a proof in his pocket of the picture he had made of his sister Ellen up a tree. "The light was very queer," he complained.

"'I think they were a little miffed.'"

"What do you think of these geese?" asked Owen, pulling a proof from his pocket.

"Evidently they felt offended," said Allan. "They seem to be turning their backs on you."

"I think they were a little miffed," admitted Owen. "Geese are mighty independent, anyway. Here's another lot I caught over by the old Dutch farmhouse, that wouldn't notice me at all."

McConnell joined them with a proof of one of his numerous attempts at Artie. This time Artie had his crossbow.

"'Over by the old Dutch farmhouse.'"

"I like that sunlight effect," said Allan.

The three boys were just entering the club rooms when Big McConnell hailed them. "Hello, Captain Kodak! What's the conspiracy now?"

"Have you seen the rooms?" asked Allan.

"No, I haven't, but I want to. I think you fellows have neglected me. I want to see what is going on the same as any one else."

"You're welcome," remarked Owen.

"This is the dark room," said Allan, indicating the dim recesses beyond the partition.

"I see," said Big McConnell. "Then this other is the light room, hey? And, oh, yes, this is the medium room over here — just half and half."

"Stop your fooling, Billy," remarked Little McConnell.

"And here I am dying to be photographed," complained Billy, "and nobody has taken me yet. It's a shame."

"Artie had his crossbow."

"We'll all take you," offered Allan, "and make a composite."

"No, you don't," said Billy. "I want to look pleasant, I don't want to look cross. You must photograph the angel side of me. I want you to take me, not to give me away," and Big McConnell roared at his own joke.

"Sit there by the window," ordered Allan.

"Yes, mister," said Big McConnell, meekly. "See that my hair is pretty and my tie straight."

Allan placed his camera on the table opposite, adjusting a box and a book to bring the camera into position. "Now, look pleasant," said Little McConnell.

Billy broke into a broad grin. "Is this pleasant enough?" he asked.

"Too pleasant," said Allan. "Look serious, please."

"I can't," cried Billy, "it's too funny. Take me as I am, or send me home to my mother."

"Well, keep still, anyway," pleaded Allan.

"Oh, I'll keep still — but can't you snap it?"

"Not in-doors," said Allan. "I must give it two seconds."

"Must you?" grinned Billy, "dear me! Two seconds! That's like a duel, isn't it? Don't they have two seconds at a duel?"

"Steady, now!" demanded Allan.

"Oh, I'm a very steady young man," protested Billy.

"There — you're taken!" said Allan.

"What, already! Why, it didn't hurt a bit. I'd never know anything had happened to me!"

"I'm going to put out a sign," Allan said; "'Painless Photography.'"

"Good idea," Big McConnell said. "'Pictures Taken Without Pain.' Everybody would come. There would be a crush. 'Line forms on this side. Walk up, ladies, and gentlemen, and kids! You'll never know what hit you.' There's millions in it!"

And Big McConnell went away with a parting warning that he wasn't one of those folks who are willing to wait very long for their proofs. "And if I don't look handsome," he said, "I'll sit again — or stand, until I'm suited."

Allan had planned several schemes for October, but the first thing that happened in October was entirely unexpected. Mr. Merring, one of the men on the *Daily Tablet*, who knew Dr. Hartel and his family, was writing an article on foot-ball for one of the magazines — he had been a great half-back himself

"'Is this pleasant enough?' asked Billy."

in his day; and he asked Allan if he would run up to New Haven with him on a Saturday to make some shots at the Yale team in practice.

"I suppose I ought to have a camera myself," said Merring, "but I've never had the time, somehow, to get at it. But you and I could work together down there."

Allan agreed to go; it was another illustration of McConnell's remark about the interesting things that happen to you when you have a camera. Merring and Allan got to New Haven at noon, and they had luncheon with two of the upper-class men, who made such a fuss over Merring that Allan concluded that Merring had been quite an important man in the university athletics.

Allan was somewhat dismayed to hear that the team would not get out to practice until nearly four o'clock. He mentioned to Merring that the light would begin to wane after three, and that with the high-pressure speed necessary to catch the rapid movements, he was afraid they could not expect good results.

"Maybe I can hurry them," said Merring; but Allan fancied that Merring did not regard the point as very important, and that he forgot the thing altogether. At all events, it was three o'clock when they started out to the field, and fully half-past three when the practice teams came out.

Allan had a full roll ready, and prepared to make the most of the situation. He and Merring took up a position opposite the middle line, and under Merring's direction he took the first line-up and several of the early plays; but Merring soon found this difficult work, as perhaps he was too much interested in the plays to care about the pictorial details. For whatever reason, he finally said to Allan, "You go ahead on your own account. Perhaps you can run in closer if I'm not with you."

Merring said that he had spoken to the Captain and the trainer, and that Allan had the privilege of the field with the camera. Allan soon found out, however, that all of the players were not familiar with his rights in the matter, for at an exciting moment

"'Chase that kid with the kodak!'"

just before a kick-off a big fellow in the line— the biggest fellow in the line, Allan thought — shouted: —

"Chase that kid with the kodak, or we'll kill him."

"Don't worry about me!" shouted Allan. "I'll get out of your way."

The big fellow paused a moment until the Captain called, "Let him alone, Barney. He's going to immortalize you in a magazine!"

Nevertheless, when a half-back came suddenly around

the end with the ball a moment later, and the whole crowd, as it seemed, after him, Allan found but one thing to do. This was to swing promptly, and leap at the top of his speed for the side line.

It was a close shave. Allan felt as if he had dodged a cyclone.

Fortunately he had some knowledge of the game, "and knew which way to run," as he told Merring afterward.

"Oh, you are a great success!" Merring said, with a laugh. "Your dodging is the feature of the day."

"An incredibly quick scattering of the players."

However, Allan found it to be impossible, with his limited experience, to get close-quarter pictures. He knew it worried the players to see him too close, and he felt that with the weakening light he could not use the highest speed of his camera shutter. Close quarters meant a blur. Distance meant small figures. Yet this was the best that could be done.

To make best use of the light, too, he shot rapidly, — and this had its natural results when he came to develop his films.

"I wish," said Merring, toward the close of the practice, "that you would catch this next play — it's a new trick the Captain is going to try; I know he has planned it, and if I shout 'Now!' you let her go as quickly as you can."

The elevens lined up again, and Allan crept as close as seemed safe.

"Fourteen — seven — twenty-one," came the voice from the tangle of legs. Allan did not hear the rest, for Merring was shouting: —

"Now!"

There was a frightful tangle of the elevens as Allan pressed the trigger, and, while his eyes still rested on the finder, there was an incredibly quick scattering of players, and five of the men, with big Barney in the midst of them, swung across his line of escape.

"Look out!" roared a voice.

Allan dropped, face down, over his camera, as he would have done over a ball.

He was prepared for the awful feet of Barney. A sound like low thunder was in his ears, he felt rather than saw a figure leap over him as he crowded close over the box; and the line had passed.

Then Merring was at his shoulder.

"Are you hurt?"

"No. I had to drop or lose the camera."

"Good play!" cried Merring. "It was a kind of touch-down!"

"And I got the picture, I think."

"Good again. Scott! Savin has made a touch-down, too!"

It was as Allan had expected; his foot-ball pictures were "undertimed," and most of those at close range were much defaced by movement of the image. But Merring was pleased, and got a number of satisfactory plates out of the batch. Allan was inclined to prize these highly whenever he thought of that thundering line.

XIV.

THE SAILING OF THE *ARABELLA*.

FOR over a week Allan was so busy over school matters that he had no time to do more than develop his foot-ball negatives; but those gay October days seemed like the best of all outing days.

"I want to take that cat-boat cruise just as soon as I can," Allan said to McConnell.

"Then you had better begin signing your crew," said McConnell.

"What berth would you like?" Allan asked.

"Well, I don't suppose you would make me first mate. I'd be satisfied with second mate. I know Owen will want to be first mate."

"I wish we three could go," mused Allan.

"Would you take the *Snorter* or the *Arabella?*"

"The *Arabella*, of course. She's quicker, and I really think she's safer. Would your mother object to your going?"

"Not with you," promptly replied McConnell. "She thinks *you're* pretty safe."

"Does she? I must try to deserve that. Anyway, I try not to take any risks when I'm off with a cat-boat. To tell you the truth, I shan't be willing to take the *Arabella* unless I can have a fellow as big as Owen along. After all, a boat like the *Arabella* is safer in every way than one of these little boats."

The *Arabella* was only twenty feet long, but she was large among the little fleet in which she moored at Kantry's dock.

"My idea," pursued Allan, "is to start Friday afternoon and make up the river as far as we can before dark, then camp inshore."

"To camp over night?" exclaimed McConnell. "That's good. I haven't camped since last summer, and that didn't count. We were right near some houses. Then what would you do next day?"

"Next day I think we might boat a good deal, make pictures, fish some if we wanted to, and get home by dark. As the tide is setting up in the after-noon now, I suppose it would be best to get down into our latitude by the middle of the day, so that if the wind weakened we should have a better chance of getting in."

"And we'll carry lots of grub," suggested McConnell.

"We shall each chip in supplies — but we are arranging the whole plan without Owen. We had better wait until we see him."

Owen came down to the club that evening. "That's a good plan," he said, when Allan found him and had set forth his programme. "But I don't think you can get the *Arabella*."

"Stowing provisions and cooking utensils on the *Arabella*."

"Why?"

"Because some one told me yesterday that Kantry had sold her."

"Now, that's too bad!" said Allan, despondently. "It doesn't seem as if any other boat could be so good as the *Arabella*."

"Why not take the *Evangeline?*"

Allan shook his head. "There wouldn't be room for us three and our cameras in that. And we never could sleep in it."

"No," Owen admitted; "we couldn't sleep in it."

"But you will go?" asked Allan.

"Yes, count on me," said Owen.

The next morning Allan went down to Kantry's before going over to the Academy, and that afternoon he met Owen with the news —

"Who do you think bought the *Arabella?*"

"Couldn't guess."

"Detective Dobbs! And I went to see him and he says we're welcome."

"To take the boat for the cruise?"

"Yes."

"That's luck."

Early on Friday morning the boys were down at the river stowing provisions and cooking utensils on the *Arabella*. Indeed, they had been down the afternoon before getting the craft into shape. They wanted to be able to lift sail and start up the river the minute they were free from school on Friday afternoon.

"You would think we were going away for a month," laughed Owen.

"Well," said Allan, "there are a good many preparations we have to make just because we are going for such a short time. We want to enjoy every bit of it when we do go."

And it certainly was with this determination that the boys made sail on Friday afternoon.

"It's good we are not old salts," remarked Owen, as the sail filled, and the *Arabella* slid into the open river, "as the whole crew would mutiny over our starting on Friday."

"Somebody told me," said McConnell, pulling at the sheet, "that Friday is a lucky day now."

"It's the best we have, anyhow," said Allan, his hand on the tiller. "I think any day is a lucky day when you can get away like this with a bright sky and plenty to eat on board, and plenty of ammunition in your cameras."

"By the way," said Owen, "I must wrap up those cameras; we might forget it."

They had carried along a large waterproof blanket in which to wrap the cameras, in case the *Arabella* shipped too much spray, and (on Mr. Wincher's advice) in which to wrap them at night, when the dampness of the river might injure the plates and the film rolls.

"We shan't take any pictures until to-morrow, anyway," said Allan.

"And suppose it should rain?" remarked McConnell.

"If it rains, we'll take some rain pictures."

"You're right," said Owen. "I think everybody takes too many sunshine pictures. It makes all photographs look alike. The painters aren't always painting sunshine."

"But I like sunshine," said McConnell, ducking his head as the boom came around.

"Oh, I don't know," mused Allan. "I've had lots of fun in the rain. The best fishing I ever had was in the rain one day."

"Oh, yes — fishing," McConnell admitted; "fishing is different. The fish like it."

"And don't you remember that ball game we finished in the rain? Wasn't it great? And the whole of that Indian Cave trip was made in the rain."

"If you like rain, you're welcome," grunted McConnell. "Sunshine is good enough for me."

"So it is for me. I'm sort of getting a waterproof on my spirits in case it does rain. Will you please notice how the *Arabella* is scooting along just now? What are you doing, Owen?"

"Getting out the feeding things."

"Already?"

"Yep. Just want to be sure things are ready."

"Owen always has a hunger on," laughed McConnell.

"You're right," confessed Owen. "Especially in a boat. Anywhere else I just have a plain appetite. But the minute I get into a boat, my stomach begins to howl for food. Besides, it's after four o'clock and I didn't eat much lunch."

"Then what do you say to a bite now?" asked Allan, "and then wait until we anchor for real supper. We must make tracks as long as the sun lasts."

"I'm with you," said Owen. "Of course McConnell doesn't want anything."

"Doesn't he, though!" chimed in McConnell. "Just watch me!"

"That's the way with these fellows that remark about other fellows' appetites," said Owen, his mouth full of biscuit. "Catch!" and Allan caught a biscuit in his left hand.

The *Arabella* was making good time in a southwest breeze, and was heading straight up the broad, majestic river. The ripples whispered under the bow, there was a chuckle in the rudder's wake, and from the throat of the boom came a grunt of contentment. The boys all shouted a greeting as they passed Mr. Goodstone in his catamaran.

"Mr. Goodstone in his catamaran."

Allan suspected that the wind would wane at sunset, and in the course of an hour turned the bow of the *Arabella* to the northwest, to which course the wind was entirely favorable. Indeed, the wind freshened, and shortly after five o'clock they were within half a mile of the western shore, which now was in shadow.

Allan then turned north again while they debated where they should anchor.

"Don't anchor yet!" pleaded McConnell.

"If we are to make a 'farthest north' to-night," said Allan, who had read "Nansen" with enthusiasm, "and do our dallying to-morrow, I think we should keep going for half an hour yet."

"Suppose we try for that cove up beyond Rodlongs," suggested Owen. "There is a spring there, and a good place to anchor."

"I remember that," said Allan; "the Canoe Club

landed there one night. But I think it will take an hour yet."

Owen thought they could do it in thirty or forty minutes, at the rate they were then going; and they

"The *Arabella* was making good time."

would have done so had not the wind fallen slightly. As it was, the *Arebella* reached the cove in three-quarters of an hour, just as the twilight began to deepen.

The boys lifted the centre-board, and pulled the bow of the boat into the mouth of a little stream that trickled from the near-by hill, and that was reënforced by the spring, to which McConnell presently started with their tin bucket.

Owen built a fire while Allan lowered and stowed the sail, braced the boom, and set about preparing for the night.

All three boys were prodigiously hungry, and Owen

worked with great zeal over his coffee, the smell of which was simply thrilling; over the bouillon, which was to be warmed; over the unpacking of the stores.

A flat stone was selected for a supper table, and in the last of the twilight, and side gleams from the fire, the boys attacked the spread with which Owen, assisted by the others, had decorated the paper covering of the stone.

"This is entirely too nice for sailors," said Owen. "We are dudes. Think of a spotless — I mean a spotted — table-cloth like this, bouillon, cold roast beef, biscuits, sweet crackers, coffee, and fresh water."

"You are spoiling us, Owen," admitted Allan. "This is too good. And to think that there is lots more left."

"Are you saving the pies for to-morrow?" asked McConnell.

"Sure," declared Owen. "Do you mean to say you want pie after all this? Pretty soon, McConnell, you'll be asking for the hard-boiled eggs we've got for breakfast."

"It seems to me I never was so hungry," said McConnell.

"Wait till the morning," said Allan. "That's when real hunger gets in its fine work."

"That's so, McConnell," said Owen. "In the morning you could eat boiled dog."

"When are we going to get up?" demanded McConnell. "Can't we have a swim, then, if it isn't too cold?"

"Of course," answered Owen, "though it's not quite correct. Sailors never swim."

"They don't?" asked McConnell.

"It seems funny," said Allan; "but they do say a great many sailors don't even know how."

"Why not?" persisted McConnell.

"Sharks, for one thing," said Owen. "Deep-water sailors get in the habit of being afraid of sharks."

"I have been thinking," said Allan, "that we had better, perhaps, draw the *Arabella* in a little farther, and let the tide leave her there. We should be floated again about five in the morning."

"Pshaw!" exclaimed McConnell. "I was hoping we could anchor."

"But if we anchored far enough out to swing with the tide, we would need to show a light."

"I really don't think we should need a light," was Owen's opinion. "It is rather shallow here, and we shouldn't need to be more than fifty feet from the shore; though we've got our lantern, and we ought to leave it, anyway, in case we get adrift. But I don't see but that we shall be better off right here out of sight, where we shall be handy to our outfit for breakfast."

"That's how it seems to me," Allan said.

McConnell was disappointed not to be able to actually sleep on the river; but he was tired, and soon began to be too sleepy to worry very much about where he was to sleep. The shore grew dark; lights gleamed on the other side of the river; the Albany and Troy night boats, with their search-lights, had passed out of sight and sound; the dark trees swayed behind them; and the crickets and locusts had begun their drowsy night chorus.

A piece of canvas, which they had brought for the purpose, was stretched to form a tent, with the boom for its central support. The blankets were unrolled and spread ; Owen lay on one side of the centre-board, Allan on the other, while McConnell completed the triangle, as he curled up across the line of their feet.

Before this had been accomplished the tide had left the boat with her keel resting in the little channel of the stream, and the *Arabella* stood almost upright. The night noises floated down from the hillside. Through the opening of the improvised tent they could see the stars.

XV.

A CHANGED SKY.

ALLAN was awakened just as dawn was breaking by a sensation of cold, and found McConnell tugging at the coverings in an effort to bury his head without uncovering his feet.

"Are you cold, McConnell?" Allan asked.

"About frozen," was McConnell's plaintive response.

"Let us get up and stir around."

They both climbed out without disturbing Owen, and soon had the breakfast fire started.

"I never knew it was so cold early in the morning," said McConnell.

Just then a head appeared from under the canvas shelter of the *Arabella*. "How about that swim, McConnell?"

"No, you don't!" retorted McConnell. "A little later in the day will do for me."

Owen laughed as he emerged from the boat, which now was afloat again. "I'm going to try for a fish,"

Owen announced; but the best he could accomplish was a very small weakfish, which he cleaned with as much satisfaction as if it had been a ten-pound bass.

Small as it was, the fish gave a delicious relish to the breakfast.

"This is simply gorgeous!" exclaimed Owen.

"Yes," said Allan, as he sipped his coffee from the tin cup, "a millionnaire in his fifty-thousand-dollar yacht couldn't live any sweller than this."

"Before we go," said Owen, "I want to make a picture of the camp. Guess I'll do it now—from that point over there," and Owen extracted his camera from the waterproof blanket in the bow, told Allan and McConnell to stay where they were, and clambered over to the view-point he had chosen. The morning was so still that at a distance of over a hundred and fifty feet they could hear the click of his shutter.

Cool as the early morning had been, the day was pleasantly mild when the sun shone, and the *Arabella* sailed away with the boys in high spirits. Allan made his course to the north again, with the wind west and freshening. They decided upon a landing at a picturesque point three miles up the river, before turning about for a leisurely journey home.

Allan and McConnell brought forth their cameras and looked them over as a huntsman might his gun, or a fisherman his rod.

"I want to make some shore pictures," said McConnell, "with long, quivering reflections in the water."

"And a white sail," added Allan, "somewhere against the green of the shore."

"And a man in a small boat in the foreground," Owen offered in supplement.

All of these elements seemed to be present at one

time or another. The shore was rich in interesting bits. The river-sailing craft gleamed in the mellow early sun. From private docks and invisible coves small boats drifted into the open. It was a fresh, buoyant morning. During the short run to the point the boys had fixed upon for another landing, the breeze became still more energetic, and the boys were delighted with the spirited way the *Arabella* behaved when Allan brought her up into the wind preparatory to landing.

With the breeze blowing inshore, they dropped anchor and landed from the stern. After all three had clambered out with their cameras, Owen and Allan went aboard again, lowered the sail, and drew a stern line to a boulder on shore.

From the point where they had landed, the river looked beautiful indeed. Ruffled by the wind, the river had no placid lines of reflection save in the turns of the shore, but the changing lines of the water under the tumbling white clouds, the smudge of New York's smoke far away to the south, the variegated river craft, coal and ice barges, tow-boats, lighters, river steamers, ferries; the gulls circling from the white of the clouds to the white of the steamers' wake — these were sights to make a boy reach for his camera now and then, until it seemed that no more plates could be devoted to the river.

They climbed to the brow of the bluff, a picturesque, wooded place, and discussed a view-point for a picture which should have a queer twist of the rocks and trees for a foreground, and for the distance the blue crest of the Palisades with the blue-green river between.

"With the breeze like this," said Allan to Owen, "I shouldn't want to try tripod work just here."

Owen had just returned from a little run overland, where he found a waterfall and an abandoned bit of orchard. Presently the three boys followed the line of the bluff to the north, and at a distance of a quarter of a mile they came upon what at first seemed like an abandoned hut, but which turned out to have for an inhabitant a queer old man, who sat just within the open door smoking a pipe.

The old man nodded to the boys and then stared past them at the river.

"I suppose he's a hermit," whispered McConnell.

"He does look rather lonesome," said Owen.

"And savage," said Allan.

"If he didn't look so savage," McConnell suggested, "I'd like to take him sitting there at the door."

Probably Owen and Allan had been thinking the same thing. Yet when they stood looking out over the river they heard a rustle in the tall grass and the queer old man had come close behind them.

"Cameras?" asked the old man in a strange voice.

"Yes," answered Allan.

"Take me," said the old man. "You will find nothing in the river so interesting as I am," and he smiled a smile so extraordinary that the boys unanimously grew uneasy.

"You are polite boys," said the old man, "and you do not say 'who are you?' Nevertheless I will tell you. I am Alexander Hamilton."

To this the boys said not a word.

"I can see your astonishment," said the old man. "You had thought with the rest of them that I was killed in the duel with Aaron Burr. Ha! ha!" and the old man laughed as a disordered phonograph

"A picturesque, wooded place."

laughs. "Yes, you thought I was killed. But I am not killed — dangerously wounded, but not killed, and I crawled away out of their sight."

Now Allan knew that the poor old fellow was crazy, but this did not make him less uneasy.

"Then you must be very old," suggested Allan.

"One hundred and thirty one this year," muttered the old man.

"The duel was in 1804, wasn't it?" asked Owen.

The old man nodded. "You remember it, then?" he added with increased interest.

"No," stammered Owen, "not exactly that — I remember reading of it."

The old man's outstretched hand pointed to the south. "On Weehawken Heights. It seems like yesterday. Ah! my dear boys, Burr was no gentleman! I wish I had time to show you my last letters to him, and my onion patch too. Do you like onions?" the old man suddenly asked McConnell.

"I like them pickled," said McConnell.

"Dear me!" exclaimed the old man, "I haven't one pickled. But suppose you photograph me anyway. I'm the oldest thing here except the hills and the river."

"Perhaps you wouldn't mind standing in the doorway of your house," ventured Owen.

"My house! Nonsense!" ejaculated the old man. "I am only living here temporarily. The place is but forty years old. I shall soon have to have another place. I outgrow them all. No; take me here with the trees. Just wait," and the old man walked away to the hut and soon returned with a straw hat on his head. "I must look like a gentleman," he said.

"You don't mind us all taking you?" asked Allan.

"No," returned the old man, "I don't mind. It is a long time since I was photographed; it was on my hundredth birthday, I think, in 1857. Dear me! how time flies when you are busy with state papers and onions and things."

"'I must look like a gentleman,' he said."

"Poor old man!" said Owen as they moved away after thanking him.

"When people talk queer like that," said McConnell, "it gives me a creepy feeling."

"I wonder how he lives," queried Allan, "and how long he has been gone that way. I wish I knew more about him."

"Well, I don't," said McConnell. "Crazy people upset me. I'm afraid of them."

"There was nothing to be afraid of," Allan insisted; "the old fellow is evidently harmless."

"Yes, I know," McConnell said; "but these harmless people — ugh!" and he shuddered. "They are worst of all."

The boys were again on the edge of the bluff. Just beyond the crest of the slope rose a shaft of rock, tufted on the top with grass, as you might fancy a stone giant with a shaggy wig. Owen made a picture from the north side, showing the shore and hills with the stone sentinel standing in the foreground.

Allan decided that if he could reach the top of the rock he could command the path by which they had come, the old man's hut, the spur of the hills, and the anchorage of the *Arabella*.

"I wouldn't risk it," said McConnell. "It looks rather narrow."

"I can do it easily," insisted Allan, "if one of you will hand up the camera afterward."

Owen took charge of the kodak while Allan, by a long reach, caught a shelf of the rock, got foot-hold, and hauled himself safely to the top.

A beautiful scene spread out before him. Low trees swayed between him and the river bank. On the opposite side was the long ledge of grass and bush-grown land and the sloping hills. North and south were the irregular lines of the shore, lighted by patches of sunlight that were moved quickly by the scurrying white clouds overhead.

"I can just see the *Arabella*," said Allan, as Owen reached far out with the camera.

To stand firmly on the head of the rock proved to be no easy matter, by reason of the narrow space and the energy of the breeze. The difficulties Allan over-came successfully as he opened his camera and set his diaphragm and shutter. It was at the moment when, with the bulb in his hand, he was sighting the camera that a huge fragment of the weather-worn rock on which he stood crumbled away, carrying with it more than half of the tuft of grass on which he stood, and Allan, after a quick effort to preserve his balance on the narrowed support, fell with the crumbling stone and disappeared from the sight of his companions.

McConnell rushed closer to the edge of the bluff with a startled cry, but he could see nothing through the fringe of leaves in the treetops below. Owen caught McConnell and pulled him back, then himself started to find a way to the river bank. McConnell started in the opposite direction, and by chance it was he who first found an opening through which he could slide and tumble to the lower level of the shore.

Tearing his way through the bushes in the direction of the shaft of rock, McConnell peered about him for some sign of Allan. When he did not find him immediately, his terror increased.

Turning farther from the foot of the rock he found the camera lying in some bushes, apparently unhurt. Then, in a little open space, he found Allan, lying on his back, his face white and still.

"Oh, Allan!" was all McConnell could say, with his heart beating so hard. The thought that Allan might be dead, stupefied him.

At the sound of Owen making his way through the bushes, McConnell sprang up and cried, "Here he is!"

The sight of McConnell's quivering lips prepared Owen for what he saw in the little opening. They knelt down beside Allan, and Owen bent closely over him, lifting his head on his arm.

"He is breathing!" cried Owen.

"Is he?" gasped McConnell; "I didn't know."

"Yes," continued Owen, peering anxiously into Allan's face; "perhaps he is only stunned. We must do something right away. If we only had some water!"

"He found Allan lying on his back, white and still."

While Owen was lifting Allan so as to place him with his head resting more comfortably, McConnell rushed to the river and filled his joined hands with water. When he had struggled back, most of the water was gone, but they sprinkled this on Allan's face and bathed his forehead.

"Do you think he has broken — anything?" asked McConnell.

"Somehow, I don't," Owen said. "It seems as if he had only knocked his head; but I can't find a cut anywhere. If we could only get him up to the queer old man's hut."

"Yes," assented McConnell. "We must do it. And I don't see how."

"You got down a shorter way," said Owen. "We must carry him that way."

But they both stared anxiously at Allan's face. *Would* he wake up again?

While they were carrying him toward the opening in the ridge by which McConnell had descended, Allan opened his eyes.

"What's the matter?" asked Allan.

Then they set him down, and McConnell began to cry and laugh at the same time, and to dance around until Owen said, "McConnell, you're as crazy as the old man of the hut."

But McConnell didn't care. He hugged Allan's hand without a word until Allan said, "What *are* you fellows doing?"

"Oh, nothing!" replied Owen. "Only trying to get you back where you started from."

Allan put his hand to his head. "Yes," he said slowly, "I slipped, didn't I?"

"I should say you did."

"And I grabbed a limb of the tree, and it broke with me, and my head struck another limb, I think, that swung me around. Yes, here it is — feel that walnut I've got here," and Owen found the spot where Allan's head had suffered in the tumble.

Allan started to his feet, then sank down again. "Oh, I'm not broken," he faintly assured Owen; "but things are swimming around frightfully. *Will* you keep still, McConnell?"

After a while Owen gave Allan a little help, and they pushed and dragged him up through the opening to the top, where the queer old man stood with his hands in his pockets.

"This way," commanded the old man, as if he knew just what had happened; and he led the way toward the hut, at the door of which he paused, made Allan sit on the step, and disappeared within.

When he reappeared, the old man had a cup in his hand. "Drink this," he said to Allan, extending his hand.

Allan hesitated. There was a dark liquid in the bottom of the cup.

"Drink it!" repeated the old man, and Allan did as he was told.

Whatever the liquid was, it made Allan feel much better, so much better that he soon began to make light of the accident and asked McConnell to go after the camera.

"These rocks are very old," said the man of the hut, "older than I am. They are getting feeble. You must not trust their strength."

"It was a close call," Owen declared fervently. "I expected to have to piece you together. But you were only out of focus and very much fogged."

"Your kodak seems to be all right," said McConnell, coming up with the camera.

Allan looked curiously at the set lever of the exposer. "I must have squeezed the bulb, anyway," he laughed. "The shutter went off."

"I wonder what sort of a thing you got," said McConnell.

"Probably some interesting sky," was Allan's opinion.

The sky! They had not noticed that within the last ten minutes the clouds to the south had grown heavier. The wind was now from the southeast and decidedly fresh.

Allan arose and felt quite steady again. "I'm all right. Good-by — Mr. Hamilton."

"Good-by," said the old man.

"I am very much obliged to you."

"You are entirely welcome, sir."

When he started to walk Allan found that he had bruised his right leg; but he sought to make light of this to himself as well as to the others; and, indeed, the stiffness which came into it while he sat on the step of the hut soon wore away.

The *Arabella* was tugging at her anchor line as if impatient to be away.

Allan looked doubtfully at the river and sky. "We shall have to put in again somewhere if things get any fresher," he said.

"I don't think it's going to be any worse," was Owen's opinion. "This is a fine breeze for a spin."

"Well, we'll try it."

After the cameras had been stowed, they debated as to whether they had better eat lunch before starting, and decided that in case they had to anchor later it would be best to spend the immediate interval in getting to a more sheltered position. "We can eat a bite on the way down," said Owen.

But when they had turned the head of the *Arabella* east to clear the point, and had made a half mile from the shore on this tack, the full force of the rising breeze became apparent, and the sky to the southeast was by no means reassuring. Allan gave the tiller to McConnell after they came about, and the two others set to work to take a reef in the sail — an undertaking to which Allan soon found that he was not equal. A peculiar weakness, the natural result of his mishap, made it imperative for him to drop on his knees and steady himself when the *Arabella* careened to starboard.

They realized now that they had been foolish in not reefing before starting, if they had not been unwise in starting at all.

"Hold her as close as you can!" called Allan to Owen, as the latter took his place in the stern after abandoning the effort to reef. "We had better make straight for shelter."

Presently it began to be plain that they could make little choice as to an anchorage. A low growl of thunder was accompanied by a spatter of rain, and in an incredibly short time the rain began to fall heavily. The wind whistled under the boom; whitecaps were all about them.

Allan and McConnell, who had drawn in on the sheet, now paid it out again, and Owen took care that the sail should not fill too full as he headed straight for the west shore.

With this precaution the boat made little headway, the sail was drenched, and its increased weight, added to the strength of the wind, kept the end of the boom much of the time in the water.

Right ahead was a shallow place and ugly rocks. To lift the centre-board here and attempt to come up into the wind would mean being blown on the rocks and the destruction of the *Arabella*.

"We must come up closer to the wind again!" cried Allan. "There is a cove a little farther to the south."

But the savageness of the wind and the wet sail made this very difficult. When they drew in the sheet, the *Arabella* took water on the starboard side. The boys were wet to the skin, and were up to their ankles in water.

"Straight for the beech tree," muttered Allan, "the water seems deep there. Don't swing her until the last minute. I'll be ready to lower away and drop anchor. McConnell, you take this other line. I'll hold the sheet free with my left."

Owen found no fault with the directions. Plainly it was the only thing to do. Owen did not put down the tiller until they were within forty feet of the shore. Then Allan let go the peak, pushed over the anchor, and they all sprang at the flapping sail.

Fortunately the cove afforded shelter from the full vigor of the wind, and made less difficult than might have been expected the task of lowering the sail. The slight shelter made it possible also to hold the lowered sail in a position to cover the pit of the boat.

"All hands to the pumps!" shouted Allan, and all three (McConnell with a drinking-cup) bailed energetically until the boat had again been made habitable. The rain fell heavily on the sail over their heads; but the situation had a pleasant flavor of adventure, and

Owen distributed rations as successfully as the cramped situation would permit.

"I wish we had something warm," said Owen. The rain had chilled them, and their clothes had no chance of drying in the present situation.

Owen finally made known his determination to get ashore and reconnoitre. He took off his shoes and socks, rolled up his trousers, and slid over the stern with a line which he fastened to the low branch of a tree that overhung the water.

When he returned, in ten minutes, it was with news of an empty old house at a short distance, a house with a fireplace where they could "get a chance to dry up." They clambered up to the old house, entered through a broken kitchen window, and soon had a blaze going in the front room fireplace. McConnell carried up the cooking traps.

"How are you feeling?" asked Owen, with a suspicious glance at Allan.

"Don't worry about me," Allan replied. "I only want to get my feet dried — and a cup of coffee," he added, glancing wistfully at the kettle.

"You'll have your coffee, Captain, in three minutes. Move up to the fire. McConnell, skin out and get me some water."

It still was raining heavily, though the wind had modified.

"If the storm keeps up," said Allan, "we shall have to spend the night here, and we might as well make ourselves as comfortable as possible." They carried up some boxes from the cellar and McConnell found an old rocker upstairs.

"The Captain, being wounded, has the rocker," declared Owen.

"They clambered up to the old house."

"The Captain doesn't want to be babied," said Allan. "The Captain will be as good as any of the crew in another hour."

The prediction almost seemed to come true. Later in the afternoon, Allan insisted on going down with Owen and McConnell to make things more secure on the *Arabella*, and to carry up the cameras, further supplies of food, and the three blankets. They couldn't reach home before dark unless with a fair wind and smooth water, and tide, wind, and water were all against them now, not to speak of the rain which continued until after dark.

Thus it happened that they passed the night in the old house, the blankets folded up for beds.

When he awoke in the morning, Allan caught sight of Owen in a far corner photographing the room and the sleepers.

"Keep still!" whispered Owen. "Let me surprise McConnell by and by, anyway."

Their breakfast exhausted the resources of the commissary department. "You see," said Owen, "we didn't expect to be away until to-day, did we?"

"We'll be late for church," chuckled McConnell.

"We may be late for supper," complained Allan. "Do you see the fog?"

"Yes," said Owen; "you could cut it with a knife. We can't budge until it lifts."

"And all the grub gone," sighed McConnell.

"We haven't even a horn," said Allan. "It makes you feel helpless. If it shouldn't clear by this afternoon, we should have to strike over to a West Shore railroad station and get around that way. I shouldn't want to worry the folks; but I haven't but half a dollar with me."

"I haven't a cent in these clothes," said Owen, as they stood looking out into the fog.

"Nor I," said McConnell.

They returned to the boat, and, to be prepared for sailing the moment the fog should lift, stowed everything on board, and drew in the stern line.

XVI.

AN UNEXPECTED VISITOR.

PRESENTLY Owen suggested that they shift their anchorage to a more advantageous point; and they had just raised the anchor when Owen exclaimed, "I've forgotten the kettle!"

"Let it go," said McConnell. It had been one of his contributions to the supplies.

"No," Owen insisted, "I don't feel like giving it up. It has been good to us and we mustn't leave it behind. Back her a little with the oar, Allan, and I'll skip up and get it."

With the oar Allan pushed the *Arabella* nearer the shore and Owen sprang out, landing on a broad stone, and disappeared among the bushes.

Dropping the oar on the deck, Allan sat down beside McConnell. The river was very still. They could see nothing but a few feet of the bank. Everywhere else was the gray, silent fog—a cold fog that made the boys shiver.

Less than a minute after he had seated himself beside McConnell, Allan felt something jar the *Arabella.* His first thought was that the boat had drifted into shallow water, and had either grounded or bumped a rock. As he turned his head he caught sight, over the bow, of a skiff, a low skiff without oars; and at the same moment the head of a man appeared above the deck line of the *Arabella.*

"Keep quiet," said the man.

The voice in which the man spoke was neither loud nor harsh, and was not above a whisper in volume; yet it gave Allan a feeling of horror. It was the voice of one exhausted, of one desperate.

"Quiet!" repeated the man, this time more threateningly, and his eyes fixed themselves on Allan in a quivering stare. As he looked more definitely into the man's face, Allan became aware that he had seen it before. Changed as the face was, there could be no doubt that it was that of the Ghost. And it arose beside Allan as the man stood up in the skiff, and, with a quick motion, stepped into the *Arabella.*

The boys now saw, with increased horror, that the man of the ghastly white face wore the clothes of a convict.

"Look here!" said the man, in the same voice, crouching beside Allan, "will you be pleasant and sociable, or must I —?" and he caught Allan by the neck with his thin hands, and struck the boy's head against the centre-board.

Allan struggled to loosen the man's hands, and then gasped, "What do you want?"

"What do I want? I want liberty. That's what I want. I want it so bad that I have been three days and three nights in this skiff, watching my chance,

"Looking out into the fog."

since I got out of there," and he pointed up the river. "They are watching and they will get me unless I can get into New York — understand me?" and the man caught Allan by the shoulder, "unless I can get into New York — into New York *with other clothes!* Do you understand? — *with other clothes.*"

"I haven't any clothes for you," stammered Allan.

"You haven't, hey? Stand up," and the man enforced his order by half lifting Allan to his feet.

At this Allan saw that, although the man had a large head, he was no taller than himself, and wasted by imprisonment, hunger, and exposure.

"No clothes, hey?" pursued the man, with something that seemed almost like a smile. "No clothes? — the very thing! Quick now, the fog helps. Quick!"

"What do you mean?" demanded Allan, who began to understand painfully well what the man did mean, and who also had begun to cast about for some plan of defence. McConnell crouched in the stern, stupefied. The *Arabella* had drifted, and the untethered skiff with it, out of sight of the shore. They were shut in by the fog.

"Quick, I tell you!" cried the man, wrenching at Allan's jacket until it had been removed. "Now the sweater and the trousers. If I have to speak again, I will speak with these," and the man shook his thin hands in Allan's face.

It was an extraordinary sight that McConnell saw, — the boy and the man exchanging clothes there in the boat; for Allan mechanically lifted the clothes the man threw from him and drew them on. On so cool a morning there was no room for debate. Excited as he was, Allan could not but foresee, with the boat adrift, that some action would soon become necessary, and the necessity for action would preclude dressing in a blanket.

"Good!" grunted the man, then he gave a start at the sound of a shout. It was Owen calling through the fog. "I see," he said, "one of you was ashore."

Allan nodded. He guessed by the sound of Owen's voice that the tide had carried them some distance, but it was impossible now to tell from which direction the voice came.

"Quick!" said the husky voice of the man, "you can join the other one. I want this boat."

"How?" faltered Allan, whose horror had been succeeded by a growing anger.

"The skiff," said the man. "There it is. If you waste a minute I'll pitch you both overboard!" As he said this, he made a stroke with the oar and soon brought the *Arabella* close to the skiff. Then he dropped the oar, clambered to the stern, McConnell making way for him, and reached for the boat with his hand. "Jump in!" he cried.

Allan had watched every movement with lips drawn, his heart beating high. To give up the *Arabella*, their cameras, and outfit without a struggle was more than he could bear. To see the man drop the oar had been a great surprise. It gave him a moment of hope, and when the man reached for the boat he saw his chance, and springing with all his force he pushed the convict over the stern into the river.

"McConnell — the sail!" he yelled, and grasping the oar gave a couple of quick splashes in the water that put the *Arabella* out of the man's reach when he had risen to the surface, spluttering and cursing.

McConnell had started forward to the lifts. Allan followed and they gave several quick hauls together, enough to lift the gaff five or six feet. The sail indicated that there was almost no wind.

Allan sprang back to the oar and called to McConnell to make the lines fast and get at the other oar. The man had climbed into the skiff and Allan saw him crouching in the bow paddling furiously with his hands — a means of propulsion which evidently he had practised in effecting his escape. His face now wore a frightful expression.

The sight of the fury in the man's eyes gave energy to the paddle stroke which Allan applied to his oar. They drew away three yards, four yards, five yards, from the skiff. McConnell's oar now joined on the port side of the *Arabella;* but the man paddled with a dreadful steadiness, fixing his upturned eyes upon them and cursing in his husky voice.

Then McConnell's foot slipped, he stumbled in the boat, and his oar went overboard. Allan made a quick reach with his own oar but could not catch the drifting blade, without turning the boat. In a few moments the convict would have the lost oar.

Again Allan sprang to the sail. "All the way up, McConnell!" he cried, and they tugged at the lines, the blood in their faces. Twice the throat of the gaff hitched; but at last the sail rose full and free, and flapped in the faint wind.

"Hold her this way!" exclaimed Allan to McConnell, and loosened the sheet.

The man had the oar. They would have known this without looking, for they could hear frantic splashing in the water. Allan added desperate strokes of his own oar to the pull of the sail. If the wind died, they were lost. The man in the skiff would have an immense advantage the moment the sail ceased to draw. Allan fancied that the convict was calculating on this chance.

Partly because of the oar, and partly because the fog left them no guide as to direction, the *Arabella* crossed the wind and the boom swung to the other side, tangling the sheet in the tiller and throwing Allan across McConnell's knees. While they struggled with the lines they lost much of their headway, and they could hear a husky yell from the man as he gained upon them. But the accident told good news. It told of a puff of wind, and when the sail had filled on the other side with the wind astern, the *Arabella* very soon led very rapidly in the race.

"We are getting away!" cried McConnell. They were the first words he had said.

The skiff and the convict grew dim in the fog.

"We have beat him!" ejaculated Allan. "He's welcome to the oar; I don't want to see him caught. But I didn't want him to take the *Arabella* — and everything."

"I was afraid we were goners when I lost the oar," said McConnell.

They strained their eyes through the fog, but could see no trace of their pursuer. Yet Allan did not feel that they were safe from him unless he could keep the wind astern, and thus be as sure as was possible that they would not cross his track in the fog.

For fifteen minutes Allan kept the *Arabella* with the wind, utterly uncertain of their direction.

All about them was the gray, still mist that filled the boys with a strange sense of mystery.

Overhead the mist was silvery, as if the sun was threatening to come through; yet when they looked on either side of the boat the veil was impenetrable.

"Did you hear something?" asked McConnell.

"No, what was it like?"

"Like a boat whistle."

Allan's face changed. Yes, he could hear the sound himself. It was distant, but he could discern the deep-throated note of a large river steamer.

"What can we do?" asked McConnell, with a new anxiety.

"I don't see what we can do. I don't know whether we are going across or down the river, and I can't tell from which direction the sound is coming."

The whistle could now be heard distinctly every few moments, and presently they decided that it was astern. "In that case," said Allan, "we are going down the river, for that must be one of the delayed night boats, and it will be best for us to keep to the west. We couldn't have gone far enough out to get into the track of the steamers." The course of the *Arabella* was turned slightly to starboard, and then the boys were thrown into new confusion by finding that the whistle was sounded on the port side. The rumble of the paddle had grown very distinct.

Allan turned the *Arabella* farther to the starboard, drawing in the sheet.

"We must make some noise — all the noise we can," said Allan; "it will be better than any signal. They would never see us in time." Thereupon Allan took two of their pans and began clashing them together as violently as he could. McConnell took two flat pieces of wood from the bottom of the boat and produced sounds like pistol shots by clapping them together. But the rumble of the paddle wheels grew louder, until Allan began straining his eyes for a sight of the approaching danger. He had never fancied it could be so difficult to tell the direction from which sound came in a fog.

"They hear us!" shouted Allan.

Several quick blasts came from the steamer whistle, the paddles turned slower, and then stopped. At the same moment the bow of a steamer seemed suddenly to grow out of nothing within a hundred feet of them, and the whistle was giving a resounding roar.

"They are passing us — it's all right!" cried Allan, with an excited laugh. Indeed, the paddles had started again.

"Now for the shore."

"Which shore?" asked McConnell.

"The west shore. We couldn't risk going across yet." Allan, with the hint offered by the wake of the steamer, turned the *Arabella* so as to head southwest. As nearly as he could guess, this was at right angles to the course he had established in getting away from the convict.

It was not until he had left the two dangers behind him that Allan began to think of the plight he was in. Then he laughed, and McConnell joined him.

"Don't you want to sit for your picture?" asked McConnell.

"No, thank you. I don't think I want to see myself in a striped suit, even for fun. I must get you to hunt up some one who will send word to Hazenfield, even if I can't go myself."

The wind drew a little stronger, and Allan began to think that the fog was lifting. It had grown sufficiently thin to justify him in running straight for shore.

"Go to the bow," Allan said to McConnell, "and yell when you see anything."

They both watched eagerly for the shore, but it was nearly ten minutes later that McConnell shouted, "A dock!"

They would have crashed into it in a few moments. Allan swung the *Arabella* and ran the boat up under the lee of the dock.

It was a small private dock adjoining a boat-house. Making fast to one of the rings, the boys climbed out.

Allan looked down at his clothes. "I wonder what any one would think of this?"

"Allan looked down at his clothes."

McConnell laughed. "You'll have to explain," he said.

The boys turned up the dock, and they had scarcely done so when a man stepped from behind the boat-house and caught Allan by the shoulder.

"No, you don't!" said the man, "no convicts here, please. If this don't beat all! Mike!" And the man shouted again, until another man came strolling from beyond the boat-house. At the sight of Allan, Mike stopped, and his jaw dropped. "Holy saints!"

"I don't mind yer gettin' away," said the first man; "but makin' use of us is too much — too much, I say."

"I'm not a convict," said Allan, "I —"

"Of course not," said Allan's custodian, "of course yer innocent. You all are."

"You don't understand," said Allan; "a man escaped and I —"

"Yes, and you couldn't resist keeping him company. Right you are, my boy, and I suppose I'd do it myself if I was in your shoes; but I'm not, and I'm goin' to keep my conscience clear. I'll hand you over and save all trouble. And yer only a kind of kid, after all."

"You're making a break," spoke up McConnell; "he's no convict. He had a fight with one, and he —"

"Now you keep quiet, young feller," said Mike. "Don't complain to us. You don't suppose we're goin' to git ourselves in a scrape, do yer?"

"What's this?" demanded a voice.

"An escaped Sing Sing man, sir," said Mike.

"A what? — dear me!" said the voice.

Allan and McConnell had started at the first sound of the voice. When they saw its owner, their suspicions were confirmed. It was Mr. Prenwood.

"Dear me! a convict!" continued Mr. Prenwood. "Why, it's not a man at all; it's only a boy —"

"Mr. Prenwood!" cried Allan, "don't you know us?"

"Know you?" stammered Mr. Prenwood, stepping closer.

"Don't you remember — Coney Island?" interposed McConnell.

"Why — upon my soul — yes, you — you are the kodak boys!"

"Yes," said Allan, "we were attacked by a convict, and he forced me to give him my clothes, and so —"

"'No, you don't!' said the man."

"And so you had to take his! Yes, yes. Blickens," said Mr. Prenwood to the man who had first encountered Allan, "you've got the shadow; the substance has escaped."

The man laughed. "I never knowed, sir."

"Come into the house," said Mr. Prenwood, who was laughing a little to himself. "Well, well! I never expected to see you this way, Mr. Allan Hartel. You see, I remember your name. And I'm glad to see you again. And what a monster this convict must have been to treat you so! Though I suppose you got off very well. Tell me all about it. Milicent," Mr. Prenwood now spoke to a lady who stood on the porch. "Don't be frightened. This is not a convict, but only a boy who was attacked by one; and these are boys I met last month at Coney Island. And I know they are hungry. Won't you get us up something nice?"

Allan expostulated that he was not hungry, that he only was anxious to get home as soon as possible, or at least to have word sent to Hazenfield.

"Blickens," called Mr. Prenwood, "get things ready on the launch. The fog is lifting, and I shall want these boys and the cat-boat towed over to Hazenfield in half an hour. Meanwhile, Allan, I'm going to get you some clothes and make you comfortable. Come upstairs and let me see if I can't fit you out while they are getting that bite ready for you. You don't look exactly right to me. Did that brute hurt you?"

Allan said the man had not hurt him, but admitted that he had not been feeling just right since he met with the accident on the bluff the day before.

"I knew it. I knew you weren't right." Allan

went on with his story while Mr. Prenwood rummaged in a closet and several trunks. "You see this is Sunday," continued Mr. Prenwood. "You couldn't go home on a Sunday without looking trim and nice. There—slip into these things. I guess that shirt will fit you."

Though they were both too greatly upset to eat much, the boys made an effort to do justice to Mr. Prenwood's hospitality, and were delighted by his cordial talk.

As he walked with them to the dock, Mr. Prenwood said he knew Owen would get home all right, somehow, and he made the boys promise to come and see him.

Blickens sat in the launch. "I'm sorry I was so rough," said Blickens.

"I didn't think you were so rough," said Allan, reassuringly, as he stowed the convict's clothes in the *Arabella*.

"A souvenir?" laughed Mr. Prenwood.

Allan explained that he had made up his mind to hand the clothes over to Detective Dobbs. "He owns the *Arabella*," said Allan, "and we can go right to his dock."

"I understand," said Mr. Prenwood. "Make the best time you can, Blickens. You see you are to take him to a detective after all."

Blickens looked rather uncomfortable. He did his utmost to make things right with Allan. The *Arabella* was made fast to the launch by a long line, the boys taking their seats in the launch.

Mr. Prenwood waved his hand, and shouted a cheery good-by as the launch and cat-boat slid out into the river.

XVII.

WINTER DAYS.

MR. PRENWOOD'S launch carried Allan and McConnell, with the *Arabella* astern, swiftly southeast to Hazenfield.

Blickens was very talkative on the journey. It was evident that he wished to atone for that which had happened at Stonyshore; and Allan felt so little resentment for what had happened that he earnestly urged Blickens to come to the Hartel house for dinner. "Then," said Allan, "I can send back these clothes by you."

But Blickens could not be persuaded to stay. "I'll wait here for the clothes, if you want me to," said Blickens at Detective Dobbs's landing; "though I don't believe Mr. Prenwood expects you to bother sending them back."

Since Blickens did not seem to be willing to accept the invitation, Allan said that he himself would return the clothes later, so that Blickens need not feel compelled to wait. Blickens said he hoped there would be no hard feelings, and shook hands good-by.

The boys were mooring the *Arabella* when Detective Dobbs came down the path with Sporty.

"Ship ahoy!" he called. "How many whales did you catch?"

"We caught a very unexpected fish," said Allan. "There's his skin," and he tossed ashore the convict's clothes.

"What do you mean?" demanded Dobbs, picking up the clothes. "This is a Sing Sing suit."

Allan hurried ashore and as quickly as he could outlined to Dobbs the meeting with the Ghost and what followed.

Dobbs uttered an exclamation of astonishment. "Of course, you don't know in what direction he went — everybody has been after him for three days."

"We left him behind in the fog; but he said he wanted to get to New York."

"But he only had one oar."

"Your other oar."

"What color was your jacket — and the sweater?"

Allan described the clothes the convict had taken, and Dobbs waited to hear no more. In six minutes Dobbs was at the station talking to police headquarters in New York. In twenty minutes every police station near the Hudson, from the Battery to Sing Sing, had a description of the clothes the Ghost wore, with information as to the skiff and the single oar. The police everywhere already had a full description of the man.

"I almost hope they don't catch him," said Allan, as he and McConnell hurried inland from the river.

"I don't know what I hope," confessed McConnell. "When I think of his knocking your head against the centre-board, I want to have him caught. When I think of how hungry he looked, and how thin his hands were, I want him to get away."

"I wonder how Owen will get home?" Allan queried. "It almost seems as if we should have tried to find him; though I'm quite sure he has walked south and will get across the river somehow."

Both McConnell and Allan found an anxious welcome awaiting them at home. The Doctor had assured Mrs. Hartel and Edith that the boys had prudently anchored when the storm came up, and that they would be home as soon as the fog lifted; yet both mother and cousin had worried greatly, and even little Ellen had made many inquiries as to why Allan did not come home.

It may be supposed that Allan's recital found highly interested listeners; that a hundred questions were asked; that some of them were answered; that Allan did not eat much dinner.

Allan insisted that he only was worrying about Owen. In the afternoon, at about three o'clock, Owen walked in to say that he had been home for a couple of hours. It appeared that after calling hopelessly into the fog, and waiting in the vicinity of the anchorage for half an hour, Owen made up his mind that the *Arabella* would have no chance of making the same point again while the fog lasted. He then turned back, and finding his way to Alexander Hamilton made inquiry of him as to the nearest way to the highroad, and was about making his way inland when a freight train on the West Shore road hove in sight. The train halted at a near-by switch, and Owen so successfully made friends with a man in the caboose that he was invited to get aboard. Three miles south he slipped off the train at Boughton, got a boat from a man he knew at the landing, and rowed across to Hazenfield.

"And so you see," said Owen, "I got out of the scrape easier than you did."

Despite the Doctor's questions, Allan continued to insist that he felt all right, that he would be all right in a little while — or the next day anyway. Yet his confidence was not justified. On the following day the Doctor betrayed by his looks that he did not find Allan to be very well. He forbade him to do any developing for a day or two longer, and kept him away from the Academy.

At the end of a week Allan was down with a fever, and the autumn colors, the stately river, the faces of his friends, the walls of the club rooms, all faded away in a troubled sleep; and other weeks passed, and there were anxious faces at his bedside, and his father would sit holding his hand and looking fixedly at him in the dim light of the sick chamber; and his head was very queer and heavy and hot, so that the ice felt like an angel's hand. And he asked them to be sure that the focus was right and that the shutter had been set, ordered McConnell to pull in on the sheet, and Owen to hand up the camera carefully.

"I tell you, mother," he said one day to Mrs. Hartel, his eyes glistening, "I've thought over the finest way to develop films! They have never thought of it! Why, it's dead easy! All you have to do is soak the film in — in — there, I've — I've forgotten just what it was, but — oh, it's very easy! I'll have great fun showing them at the club."

It was difficult to keep him from talking about cameras and expeditions and new developers.

One day he said, "It's funny that Owen doesn't get back. But I suppose he's living with Alexander Hamilton — poor old man! You had better send over and get Owen. If it hadn't been for the fog — how foggy it is again!"

Owen came every day to ask about his chum's condition; and McConnell, who was pitifully upset, never could understand why he was forbidden to see Allan, or to help take care of him.

There were many inquirers,— Major Mines, Miss Manston, Mr. Thornton, Mrs. Creigh, and other members of the Camera Club, and Detective Dobbs often called in.

It was one afternoon in late November that Allan, lying very still and quiet, with his eyes fixed on the wall at the foot of his bed, where there was a picture of the monks of St. Bernard with their dogs, said, suddenly, to his mother, —

"Did they catch the Ghost?"

At first, Mrs. Hartel thought the question was but another rambling question incident to the abating fever, and was about to utter one of the evasive replies that are offered to fevered invalids, when something in Allan's face made her understand that he was coming out of that long dream into which his mind had fallen. Then she answered him quite truthfully: —

"No. They have not found him, Allan."

"Good! I believe now that I wanted him to get away."

"But you must not worry about that."

"How many days' start has he had?"

"He has had just a month."

"A month?" Allan turned his eyes to his mother. She did not seem to be joking. "A month?"

His mother came over and stroked his hair. "Yes, a month — a long month. And my boy has had a very long sleep."

"Why didn't you wake me?" Allan asked. "I see," he said, looking into his mother's tired but cheerful eyes, "I see, I have been sick. Oh, I know it now! I can remember that the queerest things were going on in my head! One day I think it seemed to be a camera, and the lens was red hot, somehow, and somebody — who was it? — was pulling the bellows out too far. Great Scott! I thought they would break it."

"But you must not talk any more, now," urged his mother.

And he lay there for half an hour without speaking.

Then he asked: "Did Owen get back all right?"

"Yes, soon after you did."

The next day he tried for a long time to remember the wonderful plan he had dreamed of for developing films, but he could not recall the particular formula upon which the superiority of the plan rested. Perhaps, he thought, it would come back to him.

It was early in December that McConnell and Owen were permitted to come in and see him, and McConnell came in every day after that. There had been an early snow, and the boys had proofs of some snow scenes which proved to be immensely entertaining to Allan. Owen's glimpses of winter trees and snow-silvered bushes suggested many things that he himself had planned to make when winter came.

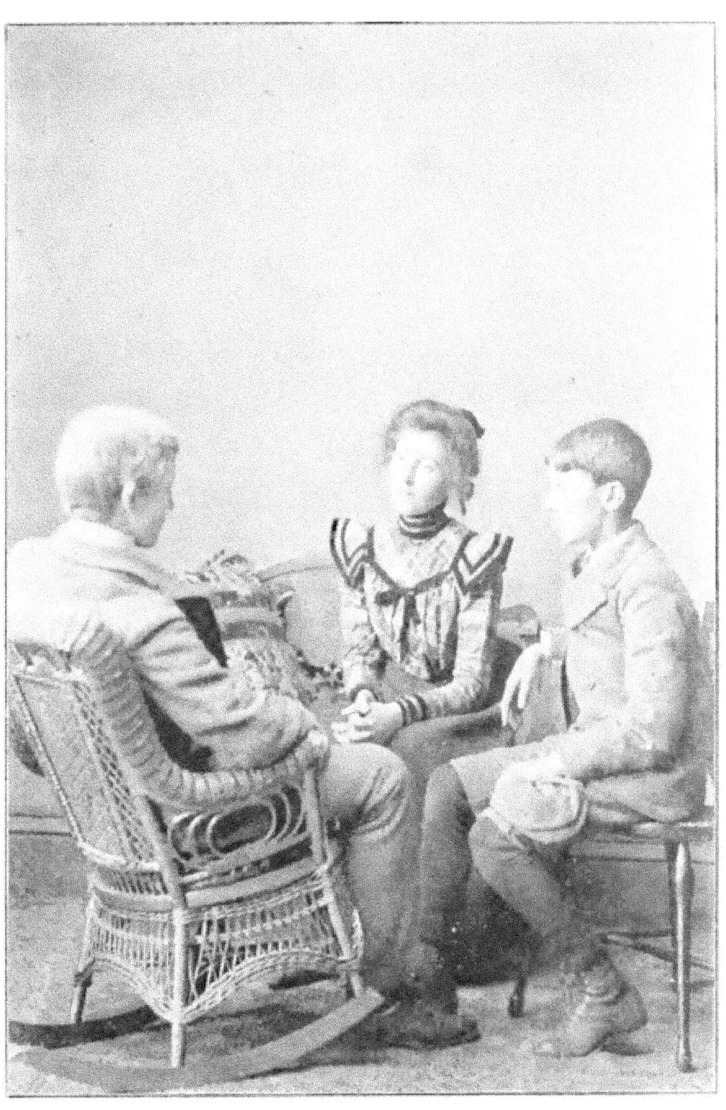

"McConnell came in every day after that."

"I shall soon be out," he told the boys, "and then I want to begin right away on some winter things." McConnell came in every day after that.

Detective Dobbs on his first visit to Allan brought a batch of pictures he had made of Sporty. "I don't have time to develop," said Dobbs, "so I let them 'do the rest' for me now. As soon as you get well," continued Dobbs, "I think the club is going to have an exhibition; they have been talking about it for some time."

Allan thought this was a capital idea. A plan for a frame big enough to hold some of his most successful pictures gave him something pleasant to think about for a whole day.

But he grew impatient to get about, and as December dragged along his resentment against the long convalescence grew deeper. It was not until Christmas Day that he came down to dinner. There were several little surprises for him, in addition to the Christmas morning surprises, that had been carried up to his room.

The dinner seemed to be in his honor, for the big cake had a frosted camera on the top, and "Captain Kodak" in fantastic letters. In the middle of the cake was a miniature imitation of a dark-room lamp with a candle burning inside.

Soon after dinner Mr. Thornton and McConnell came in with a large leather portfolio and a fine magnifying glass, which the members of the club had sent to Allan. "We all are very glad you are getting well," said Mr. Thornton. "The club will be happy to greet its President again."

"Speech!" cried McConnell.

But Allan could only say, "Thank you, Mr.

Thornton," and sit down again quickly. His head was not very strong yet, he afterward confessed.

They had a great night at the club when Allan did get back, and plans for an exhibition were talked over in earnest. There was so large an attendance of members that there was talk, too, of new club rooms, though Major Mines and Mrs. Creigh said they wouldn't give up the present quarters for the most sumptuous club outfit that could be devised.

The club exhibition took place in January, and it proved an exceedingly interesting affair. Several of the members had surprises to present — pictures which no one had been permitted to see. Among these were amusing trophies of the Central Park trip, and of the first country walk of the club. There were enlargements, beautiful bits in toned bromide paper, platinum prints, one or two gems in carbon, and to show what could be done with simple materials, Mr. Thornton had a series of "blue prints" daintily mounted.

Allan was not exactly satisfied with the framing of his prints, and wished he had had time to give more attention to that. Owen and McConnell also made a good impression. McConnell's "Water-melon Party" contained many familiar faces. Then there was Big McConnell's remarkable picture of his younger brother. Detective Dobbs exhibited his first attempts at printing. Six out of the eight prints he had framed with great pride in a gorgeous gold frame (picked out by Mrs. Dobbs) revealed the countenance of Sporty. Major Mines had some snapshots from St. Augustine and elsewhere. Miss Illwin had sent in a little landscape, which had no sooner been hung than it began to cause her great misery. "For I can see now," she said, "that I should have sent the other one."

"Big McConnell's remarkable picture of his younger brother."

"The club exhibition took place in January."

The pictures, which filled all the available space in the club rooms, were displayed for a week. And almost every night during the week there was an unusually large gathering at the rooms. On Saturday night Major Mines hung a sheet at the head of the front room, and with the aid of a small stereopticon gave an exhibition of lantern slides.

Very few of the members had tried lantern slides, but the Major's exhibition resulted in many resolutions to make slides from "pet" plates. Another result was that a few months later the club bought a stereopticon, and two nights in every month were given up to the display of lantern pictures. Allan found great enjoyment in his lantern-slide work. A

"McConnell's 'Water-melon Party.'"

device which he rigged up in the back room was soon in general use by those members who went in for slide making.

Nothing looked more beautiful on the screen than the snow pictures. The silvery tracings in the trees, the sunlight in footprints, the icicles in the summer house, the river ice pushed into pyramids in the coves, — these and a score of other themes shone with peculiar naturalness in the light of a lantern.

Allan did not go back to the Academy until February, and he had not been long at his school work again when news came that the battle ship *Maine* had been destroyed in the harbor of Havana.

XVIII.

ECHOES OF WAR.

"I SUPPOSE, Captain," said Dr. Hartel to Allan, "that if we have war —"

"If we have war!" cried Allan; "we *must* have war. They have blown up one of our battle ships!"

"Well," said the Doctor, quietly, "we don't know that yet. And I should hardly say that we must have war. War is a very serious business."

"I think we ought to make it very serious for the other side," insisted Allan, heatedly.

"It would be very serious for both sides," said the Doctor. "I was going to say that I supposed you would want to go to the front with your camera. There seems to be no doubt of that."

"I should rather go with a gun," said Allan.

"With your kodak on your shoulder?"

"You are making fun of me, father."

"I hope you don't wish me to take you seriously."

"No," interposed Edith; "no one shall take him seriously. The idea! If he talks war like this, we'll lock him up in the dark room."

"With nothing to eat but pyro and hypo," laughed Mrs. Hartel.

"Look out!" pursued the Doctor, with a twinkle in his eye, "or the Captain may indignantly resign his commission."

"Anyway," said Allan, "if I was a few years older, I'd join Company K."

The fate of the *Maine* and the prospect of war entirely changed the current of photographic enthusiasm at the club. The boys, especially, turned unanimously to war themes. Owen had made some pictures at the Brooklyn Navy Yard when he first had his camera. He had a picture of the *Boston* taking coal, and another showing the terrible twin guns in the forward turret of the *New York*. At the second lantern show in February, Major Mines displayed the mechanism of some of the rapid-fire guns, and had a capital plate showing the disappearing gun at Sandy Hook. The pictures were all quite warlike at this February meeting.

At the March meetings there came pictures of the forts in the harbor, camera sketches by Mr. Thornton from the State Camp at Peekskill the year before, and Mrs. Creigh's picture made on the deck of the *Maine*.

Then came the declaration of war, the thrill of Dewey's victory at Manila, the capture of the first sea prizes, the stir and excitement of recruiting. Big McConnell joined Company K.

Hazenfield broke out with all sorts of flags. Little McConnell boasted that he hoisted the first flag in the town; though this seemed to be a matter of debate, since Allan had swung out the Hartel flag immediately after breakfast on reading of the actual declaration of war.

"The 'sky scrapers.'"

For the next six months a large proportion of Allan's pictures had flags in them. Indeed, there was a great deal of flag in everything for a time. When Allan was in New York in July and looked up at the "sky scrapers," the sky seemed full of flags.

Of course the boys of the club lamented that they could not be at the front. The next best thing was to go to the camps; and this they did, and found the soldiers so accustomed to cameras that being photo-

A "snap" at a cruiser.

graphed seemed to cause them no uneasiness or resentment whatever.

The chief trouble, Allan found, was that he had to promise a good many prints to different soldiers who stood for him or helped him with groups and camp scenes. He had learned that promising prints is one of the most entertaining features of photography, and that keeping the promises is one of the most troublesome. However, a soldier seemed to deserve a print if he wanted it, and Allan never promised prints with more of genuine willingness than during the exciting weeks of the war.

It was at the Hempstead Camp that Allan, McConnell, and Owen found Big McConnell, who had become a corporal, and who was glad to show the boys around.

"I'm dead tired of this show," grumbled Big McConnell. "I wish they'd send us somewhere. The rations are spoiling my naturally sweet disposition. You didn't happen to see the pie-woman, did you, as you came in?"

"No," said Allan.

"I'm looking for her about now," went on Big McConnell. "I need pie, I do."

"Doesn't the cook give you pie once in a while?"

"Pie?" shouted Big McConnell. "Well, I guess not. I wish you could photograph the slush we had for dinner. You would need an orthochromatic plate and a microscopic lens."

"Did you finish the box mother sent down?" asked Little McConnell.

"Finish it?" His brother looked down in dramatic disdain. "Young fellow, I was eating the nails in the cover before eleven o'clock the next day."

"I wonder what makes you so hungry?" said Little McConnell.

The Corporal pointed across the field where the second battalion was drilling. "That, for one thing. Did you ever stop to think how heavy a musket is, and how many times its weight doubles in an hour's drill?"

Across the company street Corporal Dacey was

showing his cousin Cora how to hold a musket, and they all laughed at Cora's brave attempt to ignore the weight of the weapon.

"How to hold a musket."

Over by the Y. M. C. A. tent Big McConnell found the pie-woman, and they all went to work on pie of various denominations.

"What kind is yours?" asked Owen of the Corporal.

"I don't know," grunted Big McConnell, his mouth full. "All pie tastes alike to me. What's the Captain doing?"

This meant Allan, who had slipped off with his camera. McConnell, who went to reconnoitre, reported that Allan was taking a group of soldiers. The Corporal looked out. "The old guard," he said.

"What is the 'old guard'?" asked Owen.

"They are changing the guard," replied the Corporal. "This is the guard that has just come off duty. I shouldn't think there was light enough left for a picture."

The day had almost gone, and in half an hour the boys were on the train again on their way back to New York, where the newsboys were shouting about battles, and the papers had stupendous news eight inches high.

"I hear," said Owen, "that our smoky powder is making it awfully hard for anybody to get good bombardment pictures."

Allan thought it would be hard to report a bombardment from the ship that was doing the bombarding. "But I can't see," he said, "why they shouldn't get good shots from the boats that were looking on."

"I tell you what I'd like to have had," said McConnell, "and that was a good chance at that Matanzas mule when the shell hit him!"

"'The old guard."

Both Allan and McConnell were greatly interested in the war-ships, and read news of their doings with particular attention. When the victorious ships of Sampson's fleet returned northward, the club hired a launch and went down the river to see and picture the inspiriting naval parade. The scenes in the bay and up the river to Grant's tomb furnished material for one of the most beautiful lantern displays the club ever held.

Allan determined to make a special trip to see and

to photograph the *Oregon*, of which he was a great admirer, because "the bull-dog of the navy" seemed to embody so much of what was most American in the United States battle ships. When the *Oregon* and *Iowa* afterward came out of dry dock, and were ordered away to Manila, Allan and McConnell started off early one morning for Tompkinsville, Staten Island, off which town the war-ships were to anchor.

It happened, however, that the *Iowa* had not yet come into the bay, though the *Brooklyn* was there, and the boys recognized the graceful prow of the *Gloucester*, the plucky little converted yacht that had figured so prominently and creditably at the sinking of Cervera's fleet.

It was the *Oregon*, however, that the boys most wished to see more of, and to photograph at close quarters.

The boatmen at Tompkinsville had been charging high prices since the fighting ships had anchored off shore, and the boys were a little discouraged in their first inquiries.

In the midst of their discouragement a sailorish-looking man on one of the docks asked them if they wanted a cheap boat. This was precisely what the boys did want, and they indicated the fact to the sailorish-looking man, who thereupon lifted his finger and motioned to the boys to follow him.

The boys followed their guide for some distance and finally reached a low-roofed shop where a man with a pipe was scraping an oar. This man had a dory, but he would not rent it unless he went with it. "I'll tell you what I'll do," he said, "I'll row you around all the ships for a dollar."

"No," said Allan, "that's too much."

"The graceful prow of the *Gloucester*."

"Too much!" cried the man who had a dory, "why, I have been getting a dollar and a half a trip. What did you expect to pay?"

"I expected to pay about fifty cents an hour," admitted Allan. "And an hour was all we wanted."

"Well, I'll be losin' money," said the man; "but call it seventy-five cents."

The boys finally agreed to these terms and were soon afloat with the dory, the man pulling at the oars and asking questions about the cameras.

"Sakes alive!" he said, "but there's been camera cranks around here. It must be lots of fun, though."

There was a strong tide and the man had to bend hard at the oars while Allan and McConnell adjusted their cameras and peeped at the war-ships with the aid of their finders.

It was rather a misty morning. There was a peculiar silvery light on the water and the ships looked queerly on the shadow side next the shore. The *Gloucester* was dainty and trim. To make the glimpse of her more entertaining, a message arrived on a launch, and within five minutes the anchor had been lifted and she steamed away, evidently toward the Brooklyn Yard. The *Brooklyn*, with the shot-hole in her smoke-stack (carefully pointed out by the man of the dory), floated quietly at her moorings.

There was bustle on the *Oregon* — and it was wash-day, evidently.

"I wonder," said Allan, resentfully, "why they always have their wash out when a fellow wants to make pictures?"

The man of the dory laughed. He couldn't explain it, he said.

"How near do you want to get?" he asked.

"I want a good full-length picture," said Allan. "I think we are far enough away from her now if you will pull south a little ways."

When they had pulled south for a short distance Allan found that the three hundred feet of the *Oregon* required a long range, and the man swung the bow again to the east.

They had made several shots from their seats; but Allan now stood on the forward seat, the man slowing down again as Allan got the range.

"I'll steady you," said McConnell, bracing Allan and watching the *Oregon* as her lines swung into favorable view.

"Click!" sounded the camera, and Allan rolled the film for another shot, this time getting an almost perfect profile of the ship.

"Now, McConnell, I'll steady you."

"Oh, never mind," said McConnell, confidently; "I think I could stand better alone."

"Pull up a few strokes," Allan suggested to the oarsman, for the tide was carrying them north. "Better let me hold you, McConnell."

But McConnell, who was studying his finder, protested that he was all right. And he did make a successful shot with no more assistance than Allan's one hand at his back — assistance which he seemed to feel that he might have done without. Yet his confidence, if it had not been too great when he occupied a place on the seat, made him reckless in the moment of his success; for in stepping down his foot slipped, and the quick turn which he made to save himself sent him into the bay.

Allan sprang forward to reach McConnell, and the boatman tugged at the oars in the moment of con-

"'I'll steady you,' said McConnell."

fusion when he did not realize on which side of the stern the boy had fallen. There was a stir among the jackies in the starboard bow of the *Oregon*.

"I'm all right!" shouted McConnell.

Allan had caught an oar from the boatman's hand and now held it within McConnell's reach.

"I don't want that!" laughed McConnell, as he swam to the side of the dory. But he consented to let the boatman and Allan lift him over the side.

"Pull ashore!" said Allan to the boatman. "We must get these clothes dry somehow."

"Well, we had all we wanted everyway, didn't we?" said McConnell, trying to wring some of the water out of his clothes. "Wasn't it good that the camera dropped into the boat!"

Allan had scarcely noticed what became of the camera. He had a feeling of being responsible in some degree for McConnell's mishap, and realized that the wet clothes must be removed at the earliest possible moment, for it was a cool morning that threatened a chill.

As the boatman pulled under the bow of the *Oregon*, there was a shout from some of the jackies, and McConnell waved his hat, which the boatman had recovered with his oar.

"And so," said McConnell to his mother that afternoon, "we went to the boatman's shop and he rubbed me down with a dirty, rough towel, and put my clothes to dry by the stove while I sat on a stool with Allan's coat and an old pair of overalls on, and something like a boat sail wrapped around me. Then Allan got me a cup of tea and a sandwich from a restaurant. It was great fun; and we've got all the ships!"

Allan's report was somewhat different, naturally, but it was fully as enthusiastic in the matter of the pictures of the famous ships.

As for McConnell, he dried off so thoroughly in the boatman's shop that he suffered not a whit from his ducking.

"There's no harm in a salt-water wetting," he said, "and I got a good snap-shot at the warships, anyhow."

XIX.

RETURNED HEROES.

MEANWHILE the war had closed and the soldiers were coming home — coming home in crowded transports, in fever ships as well as in the fighting ships; coming home white and weak from the blighting tropical battle-fields.

An uproarious welcome greeted the home-coming heroes. Allan never forgot the look of Broadway, crowded with cheering thousands, when the returning volunteers of the Seventy-first started up the great thoroughfare.

It was easier to photograph the crowds than the soldiers under these circumstances, as Allan very soon found. The whole line of march was so crowded that Allan, who, during the early part of the day, had the company of Detective Dobbs, determined to strike across town and hurry up to the armory with the hope of catching the scene as the regiment reached its city home.

This proved more difficult than he had expected,

for the crowd was immense, the police could not control the lines, and the constant pressure and shifting of the throngs greatly diminished Allan's chances of keeping near the front when the regiment should arrive. A saluting gun shattered the western windows of the armory and filled the street with smoke.

At the critical moment when the regiment reached the armory, Allan, who had counted on the chance of holding his camera high enough to shoot over the shoulders of men who stood in front of him, was pushed violently to one side, and hemmed in by a standing group of men. The crowd closed about him, but he held the camera high in sheer defiance, though he caught nothing better than the jumble of heads and shoulders.

Allan had much better luck in the expedition he organized for visiting the camp at Montauk Point, on the far end of Long Island, to which the transports had been carrying the weary soldiers of many regiments. The Santiago men were there; the Rough Riders and thousands of their comrades. Big McConnell was there, and his mother found him the day after he landed. Little McConnell wanted to go at the same time, and was greatly disappointed at being left behind.

The next day Allan planned a trip for which he recruited Owen, McConnell, Joe Bassett, Philip Manton, Mrs. Creigh, and Mr. and Mrs. Austin. All save Allan and McConnell were to return the same day, and Allan's plan for staying over night was conditional on Big McConnell's valuable advice and assistance.

The day fixed for the expedition was one of those

"The look of Broadway."

that frown before they smile, and while it was frowning there were many misgivings among those who assembled for the early train into New York.

"I wonder," queried McConnell, "why the weather always tries to frighten photographers like this. We all know it will clear up." And it did.

When they reached the end of the railway journey the sun was cheering the jaded troops, whitening the sand of the beach and the canvas of the tents; glistening on the harness of the cavalry horses, on the muskets of the guard, in the folds of the regimental flags.

The great Montauk camp opened before the visitors in an imposing way; yet it seemed less of a show at the outset than the boys had expected. What it all meant came to them later. The longer they stayed, the wider and more populous it seemed to grow.

When they found the elder McConnell, he was on guard duty, and it was two hours later before he could go about with them; but meanwhile he called Terry, the big reporter, and Terry promised the boys he would help them get some pictures of interest. To begin with, he carried them off to the Rough Riders' camp.

"But, mind you," he said, "I want some prints. I can use them in a magazine article I am getting up."

The camera people were first introduced to "Teddy," the eagle mascot of the Rough Riders, who sat on the ridge-pole of a tent, and refused to pose when he was asked. But they all trained their lenses on him, and, in almost every instance, got a silhouette against the bright sky.

Allan made free to tell Terry that he would like to photograph Colonel Roosevelt.

"Well," said Terry, "they are bothering these men

to death, but I tell you what I'll do. I'll go and speak to the Colonel about something I must bother him about, and then you can improve your opportunity. You will have a good position right where you are — and so will I have a good position. I'd like a print of that, and I shall certainly put it in the middle of my mantelpiece with the inscription, 'Me and Roosevelt.'"

"'Teddy,' the eagle mascot."

Terry deliberately carried out his plan. He strolled over to the Colonel's tent, met him just as the leader of the Rough Riders was coming out, and stood there for several minutes in conversation with Colonel Roosevelt, while the camera delegation revelled in the chances afforded by their easy range.

They afterward found a group of Rough Riders who were not riding, and under Terry's direction were soon becoming acquainted with the situation of the different divisions of the camp.

"You are quite a company yourselves," said Terry to Mr. Austin, who, with Mrs. Austin and Mrs. Creigh, had rejoined the boys as they were crossing one of the roads.

"Yes," laughed Mr. Austin, "and this is Captain Kodak," he added, slapping Allan on the shoulder.

"The great Montauk camp."

"Captain Kodak," repeated Terry, who just then turned to a near-by group of men.

Allan heard Terry say to one of the men in the group, "General, Captain Kodak and his friends are within range. I warn you to preserve a pleasant and statuesque appearance."

When Allan looked toward the officer to whom Terry had spoken, he at once recognized him by the many pictures he had seen as General Wheeler.

"Who is Captain Kodak?" asked General Wheeler.

"Oh, the General has been under camera fire before," laughed one of the other men in the group.

Allan was obliged to come forward and be introduced, and General Wheeler shook hands with all of the Hazenfield delegation, and said he should not run if they insisted upon firing at him. Several pictures were taken during his short conversation with Mr. Austin.

When Big McConnell was at liberty, Allan had made arrangements with Terry to send him prints for his magazine article. Percy gazed with great pride on his big brother and hovered near him with admiring affection. Allan was scarcely less admiring of the stalwart corporal, who took the boys in tow and showed them some of the sights which they had not yet seen.

The Corporal had considered various expedients for keeping the boys at or near the camp over night. Upon consultation with Terry, it appeared that Terry's newspaper tent mate was to be away until the next day, and it was arranged that Allan and Little McConnell were to sleep in this newspaper tent with Terry. Little McConnell would rather have slept with his brother in the soldier quarters, but soon decided that staying over night in camp under any circumstances was a momentously romantic affair.

And so the evening came on, and with it all the interesting incidents of life in a camp. A long line of the Rough Riders, taking their horses to water, was visible from their tent. They heard the sunset gun and the cry of the bugles, and saw the flags come down. There were many trampings of feet in distant clouds of sand, faint shouts, and commands. The stars came out, a cool breeze drew off the sea, and the boys fell asleep.

The next day was a bustling day at the camp, an exciting and memorable day, because it was the day on which the President came. Allan and McConnell saw Mr. McKinley several times, sometimes at close quarters. The President wore a straw hat. Allan thought he looked tired and worried, though he had a pleasant, cheery word for all whom he met.

"A group of the Rough Riders."

"Terry . . . stood there in conversation with Colonel Roosevelt."

"The President wore a straw hat."

When the President went into the hospital tents, Allan and McConnell for the first time began to give close attention to these places, and began to realize more truly than before what a tragic thing war can be to those who are not hit by bullets. The thin, drawn faces of the sick soldiers made the boys' hearts heavy.

It was while the boys were standing in the shadow of one of the supply tents that two men, carrying a stretcher, halted near them, and placing the stretcher in the shadow, turned into the supply tent.

There was a movement on the stretcher, a very slight movement, and when the boys looked definitely toward its occupant, they saw a face that made their hearts leap with something like terror. At the same moment the sunken eyes that stared at them seemed to start with a responsive terror that made the ghastly face of their owner look doubly ghastly.

The man was too far gone to make a pronounced movement, but he indicated in some way that the boys were to come nearer. Allan stepped close to the stretcher.

"You know me!" whispered the man.

Allan nodded. He would have known the Ghost anywhere.

"It doesn't matter now," continued the man. "I'm done for. But you'll keep quiet, won't you?"

Allan nodded again.

"It wouldn't do you any good to give me away now. I'm sorry I did what I did to you."

Allan tried to say that he bore him no grudge at all.

"I was desperate. You understood that? And I did get away, got away and made another start. But they were after me, and I finally went where I thought they might let me alone."

The man's whisper grew very difficult to hear.

"I took care of myself for a little while. Yes, I was straight. And then one day, just as the war came, I found that they had traced me. By good luck I got a chance to enlist. That was how I dodged them again." A pitiful smile came over the man's face. "And now I'm going to escape them for good and all. No, no!" the man burst out as one of the men who had been carrying the stretcher placed his hand on Allan's shoulder. "Wait a moment!"

"You mustn't talk," said the man, firmly.

"Another word — wait!" pleaded the sick man, his face flushing for the moment. Then he whispered again to Allan. "This is the last now. You can see that — they wouldn't allow this if there was any

"'Who is Captain Kodak?' asked General Wheeler."

chance for me. I led a bad life, my boy. It was a failure. But I tried to be a good soldier. Slip your hand under here, and say, 'Good-by, Hiram Bain.'"

Allan found the man's hot hand and repeated, huskily, "Good-by, Hiram Bain."

The standing soldier's hand was on his shoulder again, and he rose up.

The Ghost's eyes seemed to be pleading for another word. Allan bent over.

"You haven't any grudge against me?"

"No," said Allan, "I haven't. I want you to get well, and to keep on — beginning over."

The man shook his head, then nodded gratefully to Allan, as if the good wish was all he wanted just then.

The two men now lifted the stretcher and moved away, Allan and McConnell staring after them until they had disappeared into one of the hospital tents.

This was the last Allan saw of the Ghost. On the following day the Ghost passed away and was buried with other soldiers who had come home to die.

When Allan and McConnell were homeward bound on the evening train, their heads full of the camp scenes, McConnell said, "I wish we hadn't seen the Ghost."

"I'm not sorry," said Allan. "I'm glad. I believe he was not so bad as they thought he was, and he did the best he could at the end. I feel better to know that he can't be hunted any more."

"Well," admitted McConnell, "I'm glad of that part, too."

A WORD AT THE END.

IT would be interesting to tell something more of Captain Kodak's experiences, but the back cover of a book is not to be ignored, and we are very close upon it.

Allan was reëlected President, and the Camera Club began to seem indispensable to the amateur photographers of Hazenfield. Many improvements were made in the dark room, and new facilities for printing were added to the smaller front room. Indeed, the certainty that the club was outgrowing the Hartel coach-house was clear to every one.

One day Mr. Prenwood came over from Stonyshore. It happened that there was a meeting that night, and as a result of Mr. Prenwood's chat he joined the club himself. This particularly pleased Allan and McConnell.

"I shall get over once a month," said Mr. Prenwood, "and see if I can't learn wisdom from you experts."

Yet to Prenwood, as to those he met at the club, perhaps the best thing to be found there was not the photographic wisdom.

www.ingramcontent.com/pod-product-compliance
Lightning Source LLC
Chambersburg PA
CBHW021513210326
41599CB00012B/1243

9783337082895